Healthy Sexual Development

James J. Neutens, Ph.D.
Director, Division of Education
Department of Obstetrics and Gynecology
The University of Tennessee Graduate School of Medicine

AGS®

American Guidance Service, Inc.
Circle Pines, Minnesota 55014-1796
800-328-2560

Grateful acknowledgment is made for permission to reprint the following copyrighted materials:

Page 17: From *The Entertainment Machine: American Show Business in the Twentieth Century* by Robert C. Toll. Copyright © 1982 by Oxford University Press, Inc. Used by permission of Oxford University Press, Inc. Page 37: From "A Gift From Mother Nature" from *Saint Raphael's Better Health*, January/February 1986. Reproduced with permission. Page 55: Excerpt from "Where Little Boys Can Play with Nail Polish" by Laura Shapiro from *Newsweek*, May 28, 1990, Newsweek, Inc. All rights reserved. Reprinted by permission. Page 64: From *The Far Side* © *Farworks, Inc.* Used with permission. All rights reserved. Page 66: Excerpted with permission of Simon & Schuster from *Zorba the Greek* by Nikos Kazantzakis. Translated by Carl Wildman. English translation Copyright © 1952, 1953 by Simon & Schuster, Inc. Copyright renewed © 1981 by Simon & Schuster, Inc. Page 71: From "Birds Nip Youth Mating Game in the Bud" by Thomas H. Maugh II, from *Los Angeles Times*, March 28, 1992. Copyright © 1999, Los Angeles Times. Reprinted by permission. Page 76: Cartoon *Zits*, "Ahem" by Jerry Scott and Jim Borgman, 10/11/98. Reprinted with special permission of King Features Syndicate. Page 80: *Foxtrot* © Bill Amend. Reprinted with permission of *Universal Press Syndicate*. All rights reserved. Page 113: Taken from the *Dear Abby* column by Abigail Van Buren. © *Universal Press Syndicate*. Reprinted with permission. All rights reserved. Page 129: Excerpt from "Slow Times at Amherst High." Copyright © 1991 by *Harper's Magazine*. All rights reserved. Reproduced from the April issue by special permission. Page 131: Excerpt "General Signs of STDs" from "STD Facts" written by Jane Hiatt with Kay Clark and Mary Nelson, Network Publications, 1986. Copyright ETR Associates, Santa Cruz, CA. Reprinted with permission.

Photos:

Front cover, upper right, ©Rob Gage; front, lower left and background, ©Jess Stock/Tony Stone Images; back cover, background, ©Jess Stock/Tony Stone Images; back, lower left, ©Cameron Hervet; p. 5, David Young-Wolff/PhotoEdit; p. 12, Bonnie Kamin/PhotoEdit; p. 19, David Young-Wolff/PhotoEdit; p. 39, Will Hart/PhotoEdit; p. 40, Myrleen Ferguson Cate/PhotoEdit; p. 43, David Young-Wolff/PhotoEdit; p. 44, Bill Aron/PhotoEdit; p. 52, top, Tom McCarthy/PhotoEdit; p. 52, bottom, Michael Newman/PhotoEdit; p. 57, Tony Freeman/PhotoEdit; p. 59, Tony Freeman/PhotoEdit; p. 62, Robert W. Ginn/PhotoEdit; p. 73, Tony Freeman/PhotoEdit; p. 74, Rhoda Sidney/PhotoEdit; pp. 95, 100, 101, 103, 104, Michael Newman/PhotoEdit; p. 115, Rhoda Sidney/PhotoEdit; p. 118, Michael Newman/PhotoEdit; p. 122, David Young-Wolff/PhotoEdit

©1999 AGS® American Guidance Service, Inc., Circle Pines, MN 55014-1796. All rights reserved, including translation. No part of this publication may be reproduced or transmitted in any form or by any means without written permission from the publisher.

Printed in the United States of America

ISBN 0-7854-1868-7

Product Number 92040

A 0 9 8 7 6 5 4 3

Contents

Chapter 1	Sex and Sexuality: A Kaleidoscope of Meanings	5
Chapter 2	Adolescence: Changes and Responsibilities	19
Chapter 3	Female and Male Roles	39
Chapter 4	Becoming Sexually Responsible	57
Chapter 5	Communication and Relationships	73
Chapter 6	Controversial Issues	95
Chapter 7	Sexual Exploitation and Harassment	115
Appendix A	Sexually Transmitted Diseases	131
Appendix B	Events in Pregnancy	136
Glossary		138
Index		141

Chapter 1
Sex and Sexuality: A Kaleidoscope of Meanings

▶ *What are the meanings of sex and sexuality?*

▶ *What are some common myths about sexuality?*

▶ *What types of sexual issues and decisions face adolescents today?*

One day at school, Emily overheard Jason and Bob talking about girls. "Hey," shouted Bob. "Did you see how she's built?"

"Sure did," snickered Jason. "She's the kind of girl I'd like to take home—but not to meet Mom."

Emily wondered if sex was all boys ever had on their minds. On her way home with her older sister Elizabeth, she asked, "What do boys think about?"

"Well," noted Elizabeth softly, "they think a lot about girls. That usually means they are thinking about sex. Sex may mean love, romance, and weddings to us, but not to them. So watch out!"

Emily was at a loss. At home and church she had been told that sex was something beautiful to be shared only by married people.

> **Sexuality**
> A personal and natural expression of a person's identity beginning at birth and continuing until death

What are the meanings of sex and sexuality?

The meanings of the word *sex* are a bit like views through a kaleidoscope—a constantly changing set of colors. Ideas about sex are constantly changing among different groups of people. When Jason and Bob talk about sex, they may discuss the way a girl is shaped or a certain sexual behavior. They don't mention love. Emily and Elizabeth feel that love, romance, and marriage are important parts of sex. Emily believes that sex is not purely physical but that it is a way of caring and sharing.

You may find that your view of sex is different from the view of your parents, friends, or dating partner. You may also find these different interpretations of sex to be confusing.

Sexuality is a word that includes more than sex. Sexuality is a personal and natural expression of a person's identity beginning at birth and ending with death. Although all of us are born sexual beings, we grow to make observations and form our own feelings about sexuality. We all come from different backgrounds. We gain different experiences.

From time to time, different things may affect the way we feel about ourselves. For instance, if Emily discovers a blemish on her forehead, she may feel unattractive. If she goes to a party later on that day, she may be shy because she feels afraid that the opposite sex will reject her. In another example, Bob may fail at some sort of athletic experience, perhaps even to the point of being publicly embarrassed. In turn, he may develop feelings that affect his ability to interact well with young women. A third example is a girl named Jill, who has had many sex partners. These partners have been careless in talking about their experiences with her. As Jill hears what they have said to others, her own feelings of self-worth may decline. She may develop some negative attitudes about sex.

A person's sexuality can bring both great joy and great confusion. If dealt with irresponsibly, sexuality can bring great pain. For a healthy understanding of your own sexuality, it helps to begin by learning correct information.

What are some common myths about sexuality?

In our modern world, we see sexual messages every day. Television, billboards, and magazine ads tell us how to dress, smell, and act. Following such "advice," claim the ads, can make us more sexually attractive. Special telephone lines and computer bulletin boards pass on sexual messages. Many adults are reluctant to discuss sexual topics in open, factual, and honest ways. As a result, much of the information young people receive comes from unreliable sources. Test your sexual knowledge by taking the Myths and Misunderstandings Quiz.

MYTHS AND MISUNDERSTANDINGS QUIZ

Listed below is a series of statements about sexuality. Read each statement carefully. Then write T if the statement is true or F if it is false.

_____ 1. Alcohol is a sexual stimulant.

_____ 2. The best, most effective contraceptive choice for all individuals is the pill.

_____ 3. Certain food substances are capable of increasing sexual desire.

_____ 4. Pregnancy cannot happen unless intercourse has taken place.

_____ 5. The larger the male penis, the more stimulated the vagina.

_____ 6. Sperm are able to cause pregnancy one to two days after entering the vagina.

_____ 7. A majority of sex crimes against children are committed by adult friends or relatives.

_____ 8. Alcohol plays a major role in many rape situations.

_____ 9. If a woman urinates after intercourse, she will prevent pregnancy.

_____ 10. Males cannot have a "real" erection until they reach puberty.

_____ 11. Masturbation can cause physical problems.

_____ 12. Regular physical activity greatly reduces sexual desire.

Sexually transmitted disease (STD)
Any disease that is spread through sexual activity

Contraceptive method
The technique used to prevent the joining of an egg and a sperm

Conception
The union of an egg and a sperm, causing pregnancy

Basing health behavior on myths or misconceptions can create serious problems with grave consequences. For example, a man had a discharge from his penis that went away by itself. Because it went away, he may have the misconception that the problem cleared up without treatment. However, he may well have a **sexually transmitted disease (STD)**—any disease that is spread through sexual activity. The disease could be spreading in his body, and he also may infect others with it. Similarly, a woman may believe that she has no chance of getting breast cancer. She may fail to do a monthly breast examination. Her faulty belief could have unfortunate results.

Now check how well you answered the quiz questions:

1. While alcohol may reduce a person's restraint, it is a drug that will slow down or even stop sexual expression. Answer is false.

2. The typical response to this question is true. However, the correct answer is false. Several issues need to be considered when a **contraceptive method**—the technique used to prevent the joining of an egg and a sperm—is selected. For example, the birth control pill does not protect against sexually transmitted diseases. The couple or individual should thoroughly explore and learn about all contraceptive techniques before any final decision is made. The pill is an inappropriate choice for many people. (See Chapter 6 for more information about the pill.)

3. No foods are known to increase sexual desire. Answer is false.

4. Many people believe that pregnancy is possible only if the penis penetrates the vagina and ejaculation occurs. However, **conception** is entirely possible if sperm have been deposited in the genital area. Conception is the union of an egg and a sperm, causing pregnancy. Also, as you will learn later, ejaculation does not have to occur for sperm to leave the penis. Answer is false.

> **Masturbation**
> *Self-touching or self-stimulation of the genitals for sexual pleasure*

5. Since the vagina conforms to the penis, the size of the penis is not important in stimulating the vagina. Answer is false.

6. The answer is true.

7. Research has shown that men are involved in about 95 percent of sexual abuse against girls and 80 percent against boys. Acquaintances, family friends, siblings, and other relatives usually commit such abuse. Answer is true.

8. According to estimates, alcohol plays a role in one-third to one-half of all rape cases. Always keep this important fact in mind. Answer is true.

9. Urine leaving the urethra of the female does not come in contact with sperm deposited in the vagina. Urination following sexual intercourse will not prevent pregnancy. Answer is false.

10. Male babies are capable of erections shortly after birth, and yes, these are "real" erections. However, it is not until puberty that a male produces sperm, thus making him capable of reproduction. Answer is false.

11. No studies have shown that **masturbation**—self-touching or self-stimulation of the genitals for sexual pleasure—causes physical problems. Nonetheless, many men and women have guilt about masturbation because of moral teachings or parental disapproval. Answer is false.

12. If this means physical activity that brings one to the point of exhaustion, this statement may be true. However, routine daily physical activity will not lower sexual desire. In fact, such activity may even increase desire in some people. Answer is false.

> ## TEENAGE PREGNANCY
> - Thirteen percent of all U.S. births are to teens.
> - By age 19, 10 percent of teenagers become pregnant.
> - Seventy-eight percent of teen pregnancies are unplanned.
> - Nearly one in four teenagers who have a baby become pregnant again within two years.
>
> ### Consequences of Early Childbearing
> - One-third of pregnant teens receive inadequate prenatal care.
> - Babies born to young mothers are more likely to be low birth weight and to have childhood health problems.
> - Seven in ten teen mothers complete high school, but they are less likely to go on to college than are women who delay childbearing.
>
> Source: The Alan Guttmacher Institute, 1998

What types of sexual issues do adolescents face today?

Every day, adolescents are confronted with sexually related health issues. The United States Public Health Service (USPHS) and the American Medical Association (AMA) have set specific goals for improving the health of America's adolescents. These goals quickly reveal the many pressing, sexually related health issues facing millions of adolescents across the United States.

Some of these health goals for adolescents follow.

GOAL 1: To reduce pregnancies among adolescents

Current statistics show that each year almost one million teenage women between 15 and 19 become pregnant. Babies of adolescent mothers are at greater risk for illness. Teenage mothers are at greater risk for school dropout, lowered income, and repeat pregnancies.

> **Human immunodeficiency virus (HIV)**
> *The pathogen that causes AIDS*
>
> **Acquired immunodeficiency syndrome (AIDS)**
> *A disorder of the immune system*

GOAL 2: To encourage postponement of sexual activity among adolescents

Sexual activity at a young age increases the risk of teenage pregnancy. It also increases the risk of having sexually transmitted diseases, including **human immunodeficiency virus (HIV)**, the pathogen that causes **acquired immunodeficiency syndrome (AIDS)**. AIDS is a disorder of the immune system.

GOAL 3: To increase abstinence from sexual activity among sexually active adolescents

Delaying sexual activity during adolescence means more than preventing pregnancy. It shows a mature attitude and an awareness that a person is still developing intellectually, socially, morally, emotionally, and physically.

GOAL 4: To increase the use of contraception that both effectively prevents pregnancy and provides barrier protection against disease among sexually active young people

Teenagers who have sex without the protection of contraceptives leave themselves open to pregnancy and sexually transmitted diseases, including AIDS. Increasing the use of contraception that prevents pregnancy and barrier protection against disease can reduce unplanned pregnancies and disease.

GOAL 5: To reduce rape and attempted rape

The current rate of rape is 370 for every 100,000 women age 12 or older. Many of these incidents occur between young people who are dating. (Rape will be discussed further in Chapter 7.)

GOAL 6: To reduce sexually transmitted diseases, including AIDS, among adolescents

Every year, three million STDs occur among people ages 13 to 19. Treatment for STDs reduces the spread of HIV. (Appendix A provides information about STDs.)

Family discussion is one of the most important factors in understanding human sexuality.

GOAL 7: To improve discussion of and education about human sexuality

An estimated two-thirds of teenagers have discussed sexuality with their parents. As the statistics on teenage pregnancy indicate, the focus many teens have on their own sexuality is narrow. They often do not consider the long-term consequences to themselves or to their babies. Many teens believe they can easily handle their sexual lives with few problems. But those teenagers who fell into the negative statistics felt the same way—that they could behave casually yet remain in control. Sexuality is a powerful life force requiring very responsible behavior. Responsible sexuality is one of life's great joys.

What do you think?

At the beginning of the chapter, you learned that Jason and Bob see sex and sexuality much differently than Elizabeth and Emily do. Think about where they each might have acquired their views about sex and sexuality. Then use the following questions to explore your own view of sex and sexuality.

1) How do your friends, parents, and peers define or view sexuality?

2) How are sex and sexuality portrayed in advertising, in music videos, and in movies?

3) How does your view of sexuality differ from all these views? How is your view the same?

4) How does your definition of sexuality guide your thoughts about sexuality and your own sexual behavior?

Think of someone you have a crush on at school, at work, in the movies, or some other place. Why does that person appeal to you? How do these appealing characteristics fit in with your definition of sexuality?

How will you decide?

Explore your own behavior in regard to each of the health goals mentioned in this chapter. For each goal, make a statement or list telling what you are doing or are not doing in regard to that goal. This list will be private.

Choose which of your behaviors you listed will help fulfill the national health goals.

Evaluate how the behaviors you chose will affect your life over the next five years. Also evaluate what your life would be like if you chose different behaviors that worked against the health goals. For instance, what if you became infected with a sexually transmitted disease or became a teen parent?

Look back and review the health issues confronting adolescents today. Decide what behaviors you will keep and what behaviors you will change.

Chapter 1 Review

Ideas to Remember

■ Sexuality is a personal and natural expression beginning at birth and continuing until death.

■ In this multimedia world, we are bombarded with unreliable sexual messages.

■ People all have a different view of sexuality as influenced by their individual background and experiences and by the people with whom they interact.

■ Myths and misunderstandings about sexuality can bring harm to ourselves and to others.

■ Your sexual decisions and behaviors have an impact on the nation's health, your own health, and the health of those in your immediate surroundings.

Questions to Ask Yourself

1) What is sexuality?

2) What have been your sources of sexual information?

3) Do you need to change or add to your sources of sexual information? If yes, how would you change them?

4) What false or mythical views of sex and sexuality do you see on television, in the movies, or in magazines?

5) Where can you go to get accurate information about sexuality?

As an individual or as a group or class, find out what community resources provide accurate sexual information to adolescents. You may want to make trips or phone calls to the library, counseling agencies, local health departments, hospitals, and special organizations offering adolescent health assistance.

Once you have done your research, construct a resource booklet containing these items:

- the names, addresses, and phone numbers of available people or organizations
- the types of issues addressed and services provided
- miscellaneous information, such as fees involved or the level of parental involvement required

Chapter 1 Review Sheet Name _____ Date _____ Period _____

Read each sentence. Write T if the statement is true or F if it is false. Rewrite each false statement to make it true.

_____ 1) The meaning of sexuality has to do with a personal expression of one's sexual identity from birth until death.

_____

_____

_____ 2) Alcohol reduces sexual inhibitions, or restraints.

_____

_____ 3) Conception is not possible unless ejaculation occurs.

_____

_____ 4) Almost one million teenage women become pregnant every year.

_____

_____ 5) A teenage mother has a smaller risk of repeat pregnancies.

_____

_____ 6) Sexual activity at an older age increases the risk of getting an STD.

_____

_____

_____ 7) One goal for adolescent health is to increase contraception that prevents pregnancy and provides a barrier against disease among sexually active young people.

_____

_____

_____ 8) The use of contraceptives helps to reduce the number of STDs.

_____

_____

Sex and Sexuality: A Kaleidoscope of Meanings   Chapter 1   15

_____ 9) Alcohol increases sexual desire.

_____

_____ 10) Few incidents of rape occur among young people who are dating.

_____
_____

_____ 11) The incidents of sexually transmitted diseases among teenagers is 3,000 annually.

_____
_____

_____ 12) Sexuality is a weak life force requiring little responsible behavior.

_____

_____ 13) Most teenagers have a broad focus on sexuality.

_____

_____ 14) In most cases, strangers commit sexual abuse.

_____
_____

_____ 15) Much of the information that young people have about sexuality comes from unreliable sources.

_____
_____

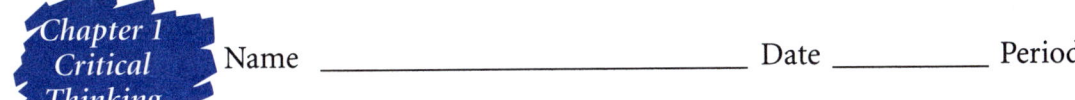

Chapter 1 Critical Thinking

Name _____ Date _____ Period _____

In the second half of the twentieth century, movies and television played a role in how we see sexuality. Portrayals of sexuality on television have changed rapidly in the past decades. Consider that in 1952, a television actress could not even say the word *pregnant*. In the excerpt below from *The Entertainment Machine*, author Robert Toll discusses how television sometimes distorts sexuality. Read it carefully. Then answer the questions that follow.

> In general although many women on television were attractive, few female stars dressed, moved, or performed in essentially sexy ways until the 1970s, the major exception being on the soap operas which carried on radio's tradition of racy daytime programming and made it even spicier. The one area in which television's evening portrayals of women even approached the veiled eroticism of 1940s' movies was advertising, which was, after all, the reason that free television existed. "The director explained that actual singing was of little import," actress Susan Barrister recalled of her audition for an automobile commercial. "It was the sensuality, the passion with which I related to the car standing before me now, that was important." In such ads, the sex appeal did not have to have anything directly to do with the product as long as it held the male viewer's interest and suggested that product somehow carried sex appeal with it. But except for some suggestive commercials, television remained very tightly censored. In 1952 Lucille Ball could not even use the word *pregnant* to tell Desi Arnaz that she was going to have a baby, and in 1960 when Jack Paar used the phrase "water closet," a European term for a toilet, the network censors deleted it from the broadcast. . . .
>
> It was in the mid-1970s . . . that [television] began to feature erotic women in action shows. . . . In 1976 *Charlie's Angels*, the story of three [attractive] young women working as private investigators, made its debut and quickly moved toward the top of the ratings. The reasons were neither the action nor the plots. "When the show was No. 3, I figured it was our acting," observed one of the Angels, Farrah Fawcett-Majors. "When we got to be No. 1, I decided it could only be because none of us wears a bra. . . ."
>
> By the late 1970s there were virtually no subjects that were taboo on television, including teenage sex, male and female prostitution, incest, adultery, rape, and homosexuality.

Sex and Sexuality: A Kaleidoscope of Meanings   Chapter 1

Consider the excerpt you have just read and remember what you learned in Chapter 1 as you answer these questions.

1) Do you think television gives an accurate picture of sexuality? Use the passage on the previous page to support your answer.

_____

_____

2) In Chapter 1, you learned some common myths about sexuality. What effect do you think television and movies have had on spreading these myths? Give an example from the passage on the previous page or from a program you have watched to support your answer.

_____

_____

3) What sexual message did the automobile commercial referred to in the passage send to viewers? What is your opinion of this message?

_____

_____

4) Reread the last sentence in the passage. Do you think this change has helped adolescents to understand sexuality or has created more confusion? Explain the reasons for your answer.

_____

_____

Chapter 2
Adolescence: Changes and Responsibilities

- What changes should adolescents expect?
- How do hormones affect reproductive organs?
- With all of these physical changes, what are your responsibilities?

Sometimes Wilson's parents treated him like a child. At other times they expected him to act like an adult. He wondered if there was something wrong with him or with his parents. He knew that his body and his emotions were going through a strange period. Compared with his friends, he was three inches taller, quite a bit heavier, and had the start of a mustache. He was looking at girls with a whole different view than ever before. Mai Lee, a friend since childhood, had become a real woman. Wilson didn't know much about what was happening to Mai Lee or himself. He just knew he was changing, and he wasn't sure into what.

While she was in junior high, Mai Lee's parents had told her she was becoming a woman. She would be able to have babies after her menstrual flow started. Her breasts had developed and her body hair had changed. Even her outlook toward boys had changed. She thought some boys acted so immature. Yet with certain other boys, she found herself excited when they accidentally touched one of her hands. She remembered her parents' comment that being a woman meant carrying more responsibilities.

What changes should adolescents expect?

Adolescence refers to the process of growing up from childhood to manhood or womanhood. The many changes that occur during adolescence involve intellectual, social, and moral development as well as physical growth. These changes bring confusing contradictions. On the one hand, adolescents are told to be responsible for their own actions. On the other hand, they must depend on the decisions of others. Adolescents are advised to delay sex until adulthood. They remain dependent on their parents economically. With all the changes and contradictions adolescents face, it is no wonder that they see this time as a big challenge. They constantly ask themselves, "Who am I?" and "Who am I becoming?"

How do you feel about your own adolescence so far?

Tremendous intellectual development occurs during adolescence. Children are concrete thinkers, meaning they see things as having a clear yes or no answer to most questions. During adolescence, this concrete way of thinking gradually gives way to more abstract or theoretical thinking. The adolescent comes to realize that many different factors can form the right answers or decisions. As a teenager, you are able to think logically and to solve problems through deduction. This means that given certain information, you can draw conclusions from it. For instance, you may conclude that if you become sexually active, you will have to think about pregnancy, contraception, and even about getting a disease such as AIDS. Keep in mind, though, that this change in intellect takes time, experience, knowledge, and patience. Most adolescents are still developing intellectually.

Social changes also occur during adolescence. As you make the transition through adolescence to adulthood, you will rely less and less on your parents and family and more on friends and peers. Through your peers, you are likely to be faced with several issues. These may include driving recklessly, using alcohol and other drugs, and participating in risky sexual behavior. Seeking your own identity is a big challenge because so many different people are telling you what to say and how to act. Your friends may tell you, "Everyone is doing it." Your parents will probably tell you, "You shouldn't be doing it."

Sitting in the middle, you must face these issues and decide for yourself.

Your ability to make decisions also relates to your moral development. Morality refers to "right" and "wrong" behavior. In early moral development, we are taught that wrong behavior brings punishment, whereas right behavior brings rewards. Small children begin to develop moral judgment by realizing that the right choice may bring a treat. The wrong choice may mean punishment.

The next stage of moral development involves doing what is right according to those in authority or acting simply to satisfy our peers and friends. In the final stage of moral development, we become aware of the rights, needs, and feelings of ourselves and those around us. Also, we gradually develop moral principles that guide our behavior. We learn to apply our own rules to the different situations we face.

WHAT ARE YOUR RESPONSIBILITIES?

As you unfold during adolescence, you are responsible for setting the direction your life takes. It is now your responsibility to become "somebody." According to psychologist Yetta Bernhard, becoming somebody means to:

- Choose and focus on a goal. (This involves choosing one short-term goal that will help lead to a long-term goal.)
- Make a commitment to achieve your goal.
- Decide if your goal is practical, through time and experience.
- Learn from mistakes along the path to the goal.
- Make new decisions and develop new plans if changes are needed.
- Be patient with yourself, especially if some plan fails.
- Accept some pain and discomfort as a common part of life.
- Believe in your own worth, even though some things may not be perfect.
- Never give up on your potential.

Hypothalamus
An area in the brain that directs the pituitary gland to release certain hormones during puberty

Pituitary gland
A small gland in the brain that releases hormones that influence other glands

Endocrine system
The body system consisting of ductless glands that release hormones

Testosterone
The male sex hormone

During our intellectual, social, and moral development, we become more independent while at the same time learning to respect the independence and rights of others. As we learn about ourselves, it is normal to shift back and forth between the desire to conform and the ability to choose to be ourselves. You will learn that you can be yourself and still be sensitive to others.

What about changes in physical development?

As people move through adolescence, they experience two stages of physical development. These stages are the growth spurt and puberty. The growth spurt is a sudden increase in height and weight. For females, it usually begins around age 12. It may begin as early as age 8 or as late as 17. The average age for the growth spurt in males is 14. It can occur as early as age 11 or as late as 18. Not all parts of the body grow at the same rate during this time, so adolescents may feel awkward. Also, no two people develop at the same rate or in the same way.

Puberty refers to changes in the body that make people physically capable of reproduction. A small area in the brain called the **hypothalamus** prompts both the growth spurt and puberty. The hypothalamus affects the **pituitary gland**, sometimes called the master gland, which then releases hormones to many other glands. As a result, people experience changes in the brain, bones, muscles, skin, and sexual organs.

How do hormones affect reproductive organs?

Hormones are a part of a much larger body system called the **endocrine system**. The endocrine system is the body system consisting of ductless glands. Hormones are the chemicals that these endocrine glands release. Hormones are chemical messengers that stimulate and influence body processes.

The pituitary gland secretes or sends out important hormones. Some hormones bring about changes in height, weight, and muscle development. Others stimulate changes in the reproductive organs. These others help produce **testosterone** and sperm cells in males. In females, the

> **Estrogen**
> *A female sex hormone*

hormones stimulate the production of **estrogen** and the development of eggs.

Both male and female sex hormones are produced in the testes in the male body. Both male and female sex hormones are produced in the ovaries in the female body. A boy's body has more male hormones. A girl's body has more female hormones. The sex hormones produced by the ovaries or testes cause the development of secondary sex characteristics.

What are secondary sex characteristics?

During puberty, primary sex characteristics, which are the male and female reproductive systems, undergo tremendous changes. Circulating hormones cause a general increase in the size, shape, and physical maturity of the body. Secondary sex characteristics also develop during this time. Secondary sex characteristics include changes in height and weight as well as changes in body hair, skin, and sexual desire. Breast development, menstruation, and ovulation occur in females. Changes in sperm production and ejaculation and a dramatic change in the voice occur in males.

Skin When hormones begin to circulate in the bloodstream, they may cause oil glands to increase their secretions. Oil can in turn cause acne—pimples on the face, neck, chest, or back. Acne is usually a temporary condition that is hard to treat, although good hygiene and some skin preparations may help.

Sweat glands Another secondary sex characteristic is changes in the sweat glands, especially under the arms. Odor results when perspiration contacts bacteria on the skin. Again, basic good hygiene keeps perspiration odor under control. In addition, deodorant products, which were unnecessary before puberty, help to cover the odor.

Breast development For some females, breast development may occur as early as age 8, while for others it may occur as late as age 13. The breasts do not grow at the same rate, so one may be larger than the other. Breast size or shape has no bearing on sexual response or on nursing a baby. Some males may experience a slight increase in breast size and have tenderness in the nipples. This is normal and often will go away in a year.

Body hair Pubic hair, or hair around the genitals, is usually the second sign of puberty for females. It generally grows in before underarm hair does or before leg hair coarsens. For males, pubic hair, underarm hair, and facial hair all grow at puberty. Some males grow a great deal of hair on their chest, arms, legs, and back. Another change for some males is that the hairline on the forehead begins to recede. With both males and females, the amount and type of body hair depends on genetic background.

Voice In males, the voice box, or larynx, gets bigger, causing the voice to deepen. During this change, a young man's voice often "cracks," going quickly from low to high or high to low pitch. In girls, the larynx also changes and the voice becomes deeper and more mature.

Sexual desire Release of the sex hormones causes both males and females to respond to sexual aspects of a movie, a picture, or a person. They may respond to almost anything that they find appealing. This response may include a faster heartbeat, quicker breathing, an excited feeling, and clammy hands. In the male, it may cause an erection, or hardening of the penis. In the female, it may cause a wet feeling in the vagina. These are normal responses that occur naturally. While you may not have control over how you respond to some stimuli, you do have control over what you do about your responses.

What about female sex organs and menstruation?

Perhaps the best way to understand the changes that a young woman goes through is to study a few illustrations. Figure 2.1 shows the female organs and the pathway of a fertilized egg. Figure 2.2 on page 26 shows the female reproductive system. Look at these figures while reading this section to help your understanding.

Ovaries A female has two ovaries, or sex glands, one on each side of the uterus. The ovaries are necessary to produce the female sex hormones and to release egg cells. After a female

Fallopian tubes
Two tubes, one extending from each ovary to the uterus, in which released eggs travel

reaches puberty, usually one egg matures each month. An ovary releases an egg, which travels into the fallopian tube. Usually the ovaries alternate releasing an egg. This release is called ovulation.

Fallopian tubes A woman has two **fallopian tubes**, one for each ovary. Each fallopian tube is flared out on the end, somewhat like the bell of a trumpet, and extends from the ovary to the uterus. These tubes are about three to five inches long and serve as passageways for released eggs. After ovulation, the egg makes its way into the fallopian tube where it travels to the uterus. Conception usually takes place in one of the fallopian tubes.

Uterus The uterus, sometimes called the womb, is a hollow structure with three linings, one of which is a muscular lining. It is shaped somewhat like an upside-down pear. The uterus is about two inches wide across the top, one inch wide across

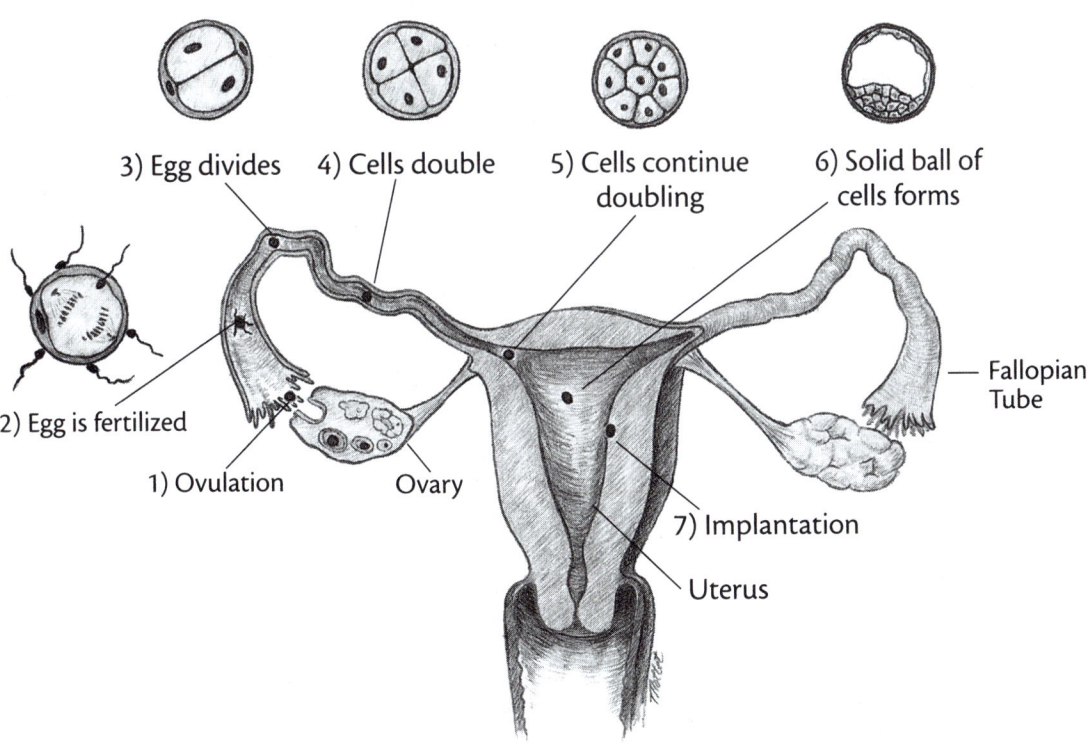

Figure 2.1. Movement of a fertilized egg from ovulation through implantation

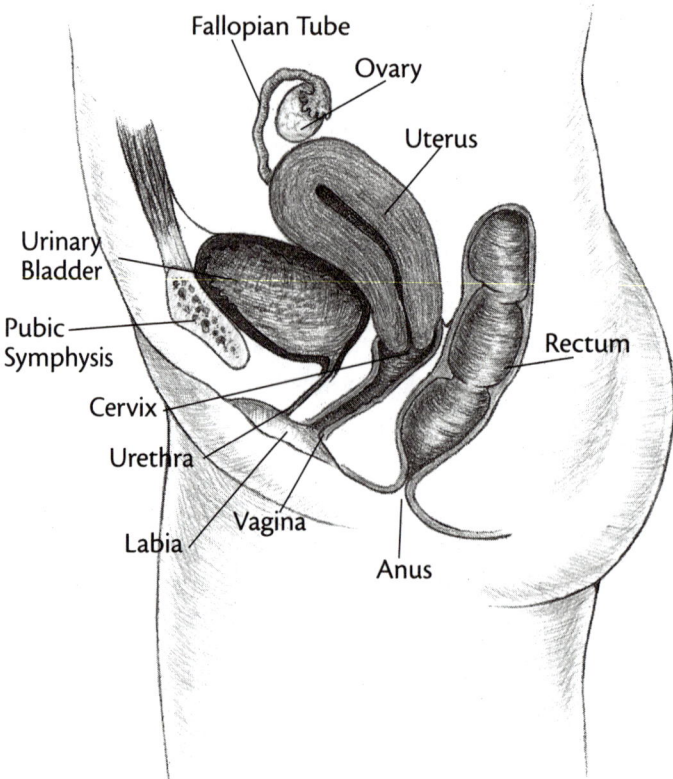

Figure 2.2. The female reproductive system

the bottom, and three inches in length. Each month the inner lining of the uterus thickens to get ready to receive an egg that a sperm has penetrated. When the fertilized egg reaches this thickened lining of the uterus, it will become implanted to remain and grow for nine months. The uterus stretches to contain an eight- or nine-pound baby. You can see that the uterus is a remarkable, stretchable organ. (Appendix B, which describes the events of pregnancy, shows this expansion of the uterus.) After a birth, the uterus shrinks down to its usual size. If fertilization does not occur, the egg is not implanted. The unfertilized egg passes out of the uterus through the vagina along with much of the thickened lining. This breakdown and removal is called menstruation. The muscular lining of the uterus, which is important in pushing the baby out at birth, is not removed during menstruation.

Cervix
The narrowed part, or neck, of the uterus

Hymen
A thin membrane inside the vaginal opening

Labia
Two folds of skin surrounding the vaginal opening

Clitoris
A tiny organ at the top of the innermost labia containing many nerve endings

The **cervix** is the narrowed part, or neck, of the uterus. If you touch the end of your nose, you will know about how firm the cervix feels. The cervix is important in keeping the baby in the uterus while it grows. During the birth process, the cervix opens up, or dilates, to let the baby out.

Vagina The vagina, sometimes called the birth canal, extends from the cervix to the outer female sexual organs, or genitals. The vagina serves as a passageway for (1) menstrual flow, (2) sperm during sexual intercourse, and (3) the baby during birth. The vagina has a mucous layer for lubrication as well as muscle and elastic layers. Usually the walls of the vagina are almost touching. However, they are capable of expansion so that a baby can be delivered. Similarly, the vagina expands and contracts to fit a man's penis during sexual intercourse.

Vaginal opening In young girls, a thin membrane called the **hymen** may be present just inside the vaginal opening. The hymen is not a solid membrane, since an opening must be present to let out menstrual flow. While sexual intercourse may break the hymen, it can be broken in many ways such as horseback riding, bicycling, or inserting a tampon. Usually the hymen is open and flexible enough to allow for the insertion of a tampon. Some girls are born without a hymen.

Vaginal lips Two folds of skin surround the vaginal opening. These are called vaginal lips, or **labia**. The labia protect the vagina from germs. Also, they have many blood vessels and nerves that make them sensitive to sexual stimuli. The innermost lips form a small hood at the top to cover the **clitoris**. This female organ has as many nerve endings as the penis and also many blood vessels. While the clitoris has no major role in reproduction, it is important for being sensitive to sexual stimuli.

The menstrual cycle is the five phases that begin in a female at the start of sexual maturation. The menstrual cycle occurs about every 28 days thereafter. The box on the next page describes the five phases of the menstrual cycle and Figure 2.3 illustrates them.

PHASES OF THE MENSTRUAL CYCLE

Pre-Ovulatory Phase The pre-ovulatory phase is the time when a new egg inside the ovary begins to mature. At the same time, the lining of the uterus thickens to prepare for implantation of a fertilized egg. The hormone estrogen causes the lining to build up.

Ovulation The release of a mature egg from the ovary is called ovulation.

Post-Ovulatory Phase The post-ovulatory phase occurs immediately after ovulation. Hormones secreted by the ovary help keep the thickened lining of the uterus in place to receive a fertilized egg.

Premenstrual Phase This phase takes place if the egg is not fertilized. The thickened inner lining of the uterus begins to break down in preparation for being expelled from the uterus.

Menstruation (Menses) Menstruation, which begins the menstrual cycle, occurs every 28 days or so. At that time, the thickened inner lining and unfertilized egg flow from the uterus and down through the vagina.

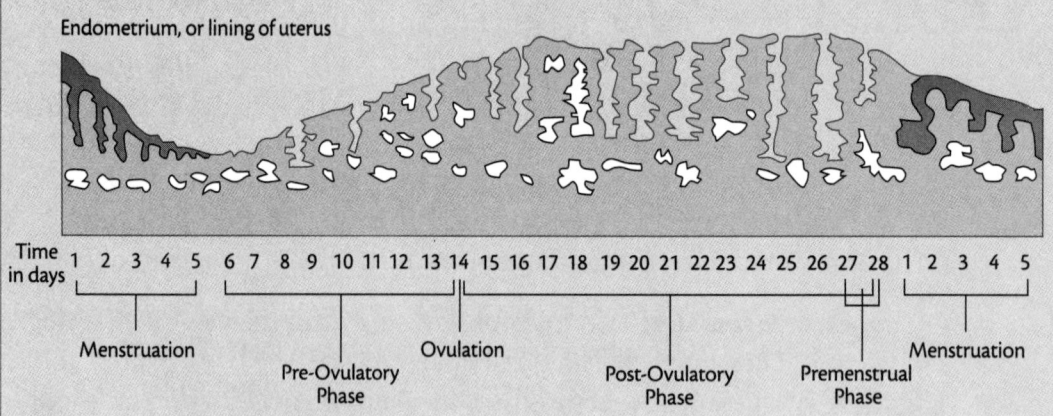

Figure 2.3. The menstrual cycle The numbers are a guide to show about where the five phases take place. The menstrual cycle may range from as few as 20 days to as many as 42 days.

What happens to the male sex organs?

As with the female's sexual organs, puberty brings about physical changes that allow the male to father a baby. Figure 2.4 shows the organs of the male reproductive system and the pathway of the sperm. Refer to the figure while reading this section.

28 *Chapter 2 Changes and Responsibilities*

Scrotum
The sac-like structure holding the testes

Seminal vesicle
One of two glandular structures that secretes seminal fluid during ejaculation

Seminal fluid
A fluid secreted by the seminal vesicles to mix with sperm

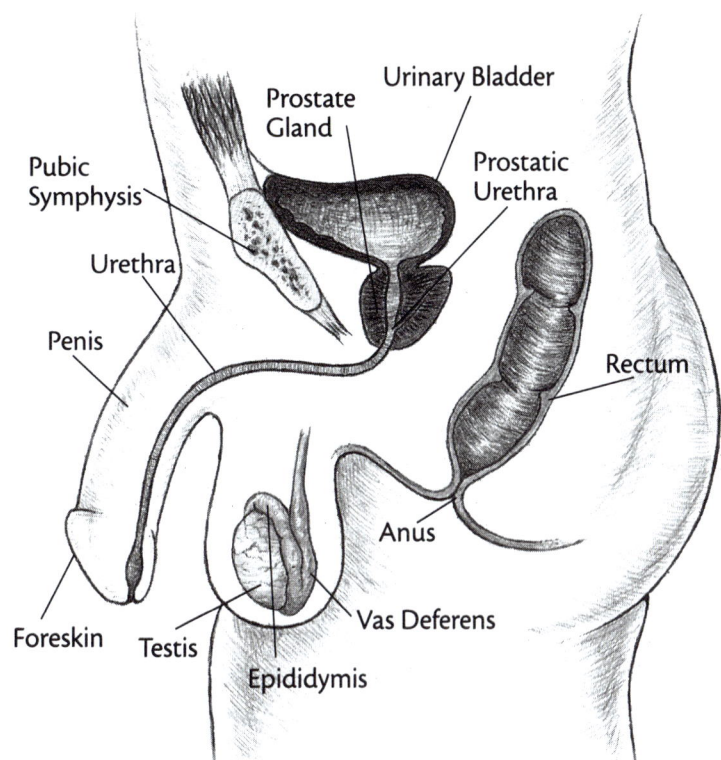

Figure 2.4. The male reproductive system

Scrotum The **scrotum** is the sac-like structure that holds the testes. The scrotum also plays an important role in regulating the temperature of the testes. For sperm to be produced, the testes must be a few degrees lower than normal body temperature. The muscles in the scrotum pull the testes closer to the body when the testes are cold. The muscles allow the testes to descend away from the body when they are hot. This helps keep the temperature of the testes constant.

Testes The function of the two testes, or testicles, is to produce sperm and the male hormone testosterone. The sperm are actually produced inside tiny tubes within each testis. Millions of sperm are produced in the tiny tubes each day.

Seminal vesicles The **seminal vesicles** are glandular structures that adjoin tubes from the testes. During ejaculation, the seminal vesicles add fluid, called **seminal fluid**, to the sperm for passage through the urethra.

Prostate gland
A small, muscular gland that produces a neutralizing fluid that protects sperm and aids in sperm movement

Foreskin
A fold of skin covering the head of the penis

Circumcision
A surgical procedure involving removal of the foreskin from the head of the penis

Urethra
The passageway from the bladder to the opening of the penis through which urine and semen pass, each at different times

Ejaculatory duct
A duct extending through the center of the prostate gland to the urethra

Nocturnal orgasm
A person's sexual arousal and response that occurs during sleep

Prostate gland The **prostate gland** is a small muscular gland that produces a fluid important to sperm movement. This fluid also helps neutralize the acid environment of the vagina. Sperm unprotected by this fluid would perish in an acid environment. The prostate gland secretes this fluid, which makes up the greatest part of the semen.

Penis The penis actually consists of three tubes that contain the sponge-like tissue necessary for an erection. An erection takes place when blood flows into and fills the sponge-like tissue, thereby making the penis larger and longer. All males are born with a fold of skin called the **foreskin** covering the head of the penis. In some males, the foreskin is removed shortly after birth through a surgical procedure called **circumcision**. Circumcision may be performed because of religious customs or for hygienic reasons.

Some males worry about penis size. This is a needless worry, since penis size has nothing to do with the ability to please a woman sexually or with the ability to reproduce. The vagina conforms to the size of a man's penis during sexual intercourse.

Urethra The **urethra** serves as a passageway from the bladder, through the prostate gland and all the way down through to the opening of the penis. (Follow this path on Figure 2.4.) The urethra carries both urine and sperm, but not at the same time. To keep any urine from escaping during ejaculation, the ejaculatory duct closes off the bladder exit in the prostate gland. The **ejaculatory duct** extends through the center of the prostate gland to the urethra.

One event that occurs in both males and females is called a **nocturnal orgasm**. A nocturnal orgasm is a person's sexual arousal and response that occurs during sleep. Nocturnal orgasms are often referred to as "wet" dreams, since males ejaculate semen. Usually sexual dreams accompany nocturnal orgasms, which are normal and simply may be a sexual release for the body.

With all of these physical changes, what are your responsibilities?

What are your responsibilities as an adolescent in regard to all the changes that are taking place in your life?

It is important for young men and women to be responsible for maintaining a healthy reproductive system. For women, this means practicing two preventive health behaviors. The first one is a monthly breast self-examination for the early detection of breast cancer. The Action for Health on page 34 shows how this can be done. Women of all ages are encouraged to make this exam part of their life. It helps to do it in the same place, at the same time of day, and at the same time during the menstrual cycle.

The second preventive behavior for women is to schedule a yearly pelvic examination with a physician. This examination and a test called a Pap smear can detect early signs of cervical cancer and other problems of the reproductive system. If a woman experiences irritation in the vagina, she should consult her doctor promptly rather than wait until her yearly exam.

Men should perform a monthly testicular self-examination to detect testicular cancer as in Figure 2.5. The second preventive behavior for men is to schedule a yearly prostate examination with a physician. The importance of this examination increases as a man reaches age 35.

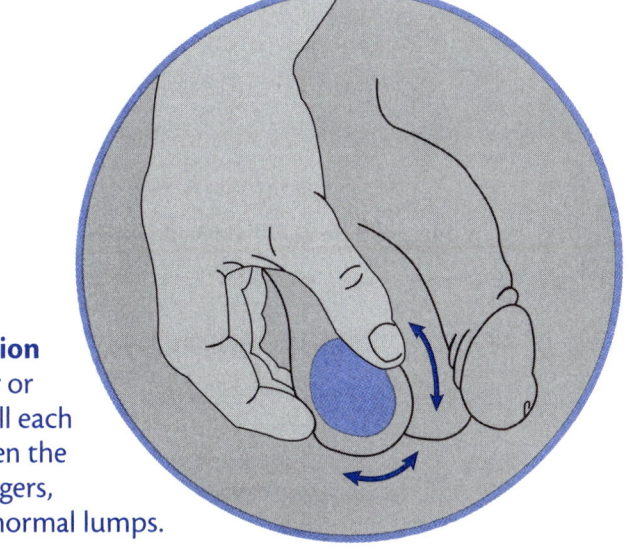

Figure 2.5. Testicular self-examination After a shower or bath, gently roll each testicle between the thumb and fingers, noting any abnormal lumps.

What do you think?

Reflect on the physical changes that occur during puberty. These changes can be embarrassing, exciting, confusing, and fascinating. A number of adjectives can be used to describe them. Think about the changes that have happened and have yet to happen to you.

1) How have the changes of puberty affected you?

2) Do you think your body is likely to change any more? If so, in what way will it change?

3) How might future changes affect you? How can you prepare for these changes?

4) How has the rate of your physical changes compared with that of your two closest friends?

5) Imagine you are near the end of adolescence. What would you say to help someone else going through the changes of adolescence?

How will you decide?

Explore the many activities or behaviors you have participated in with your friends and peers. In ten minutes, list as many of these activities and behaviors as you can. Include without judgment activities that you have enjoyed, have not enjoyed, and may have been urged to do.

Choose the two activities or behaviors you feel most comfortable doing and the two you feel least comfortable doing and circle them. Put a star by the activity that was most enjoyable. Cross out the activity that was least enjoyable.

Evaluate why you feel more or less comfortable participating in certain activities. Do the activities you feel comfortable with involve your skills and abilities? Do the activities you feel uncomfortable with go against your values, preferences, or what you've been taught?

Chapter 2 Review

Ideas to Remember

■ Adolescence is the process and period of growing from childhood to adulthood.

■ Social changes during adolescence involve a switch from parental guidance to peer influence and eventually to making decisions independently.

■ The growth spurt and puberty are the two major stages of adolescence.

■ Moral development involves a gradual shift from seeking rewards and avoiding punishment to conforming with peers and the wishes of authorities. It finally shifts to respecting others and using broad moral principles for guidance.

■ Puberty is changes in the body that make a person physically capable of reproduction.

■ The endocrine system causes many of the physical changes during adolescence.

■ An increase in sexual desire and response to sexual stimuli is natural after puberty. However, you are responsible for and can control what you do about those feelings.

■ An adolescent must accept several new responsibilities as he or she grows into adulthood.

Questions to Ask Yourself

1) What is adolescence?

2) How far along do you feel you are in your own intellectual, social, moral, and physical development?

3) What are the major phases of the menstrual cycle and in what order do they occur?

4) List five body parts of the female reproductive system and the functions of each.

5) List five body parts of the male reproductive system and the functions of each.

PRACTICING A NEW RESPONSIBILITY

Reread the section on page 31 about your responsibility to maintain good physical health. Find a private time and place at home to carry out the appropriate self-examination. If you are doing a breast self-exam, follow the directions provided in the illustrations on this page. If you find a lump, make an appointment with your doctor. Look ahead on a calendar and mark the next day that you will repeat your exam.

Breast Self-Examination
Before a mirror: Check for any changes in the shape or look of your breasts. Note any skin or nipple changes such as dimpling or nipple discharge. Inspect your breasts in four steps: arms at sides, arms overhead (Figure 2.6), hands on hips pressing firmly to flex chest muscles (Figure 2.7), and bending forward.

Figure 2.6

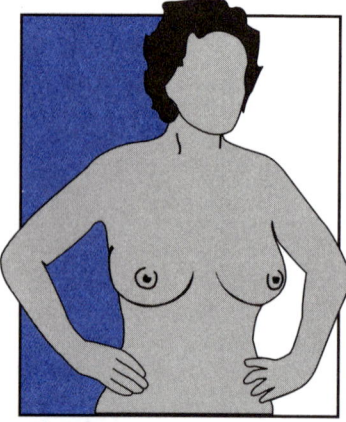

Figure 2.7

Lying down: Place a pillow under your right shoulder. Put your right hand under your head. Check your entire breast area with finger pads of your left hand (Figure 2.8). Use small circles and follow an up and down pattern (Figure 2.9). Use light, medium, and firm pressure over each area of your breast. Repeat steps on your other breast.

In the shower: Raise your arm. With soapy fingers, use your other hand to check your breast (Figure 2.10). Use the method described for the lying down examination. Repeat on the other breast.

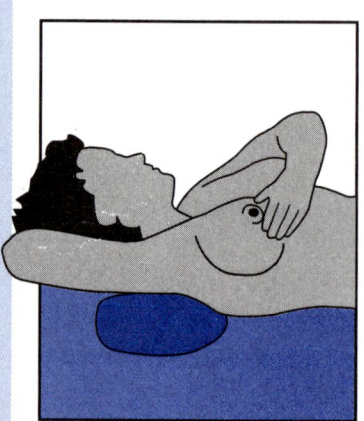

Figure 2.8

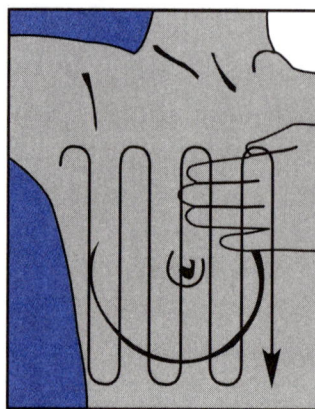

Figure 2.9

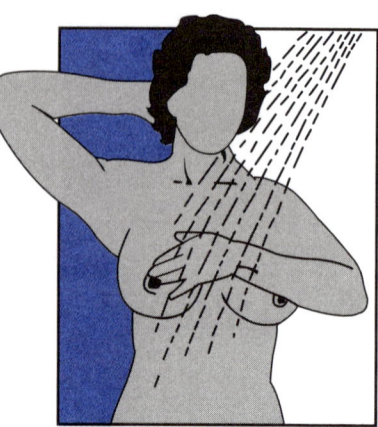

Figure 2.10

34 Chapter 2 Changes and Responsibilities

Chapter 2 Review Sheet

Name _____ Date _____ Period _____

Part A Match the description in the left column with the correct item in the right column. Write the letter of the correct answer on the blank.

_____ 1) Function is to produce sperm and testosterone

_____ 2) Test to detect early signs of cervical cancer

_____ 3) Principles that guide our behavior

_____ 4) Changes in the body that make you capable of reproduction

_____ 5) Gland that secretes hormones that produce changes in the sexual organs

_____ 6) Chemical messengers that stimulate and influence body processes

_____ 7) Consists of ductless glands

_____ 8) Male sex hormone

_____ 9) Female sex hormone

_____ 10) Physical attributes that form at puberty

_____ 11) Female sex glands that produce hormones and release eggs

_____ 12) Union of the egg and sperm to bring about pregnancy

_____ 13) Five phases that begin in a female at the start of sexual maturation

_____ 14) Two folds of skin surrounding the vaginal opening

_____ 15) Female organ that is an important receiver of sexual stimuli

a) endocrine system
b) puberty
c) ovaries
d) estrogen
e) testes
f) menstrual cycle
g) pituitary
h) labia
i) hormones
j) Pap smear
k) secondary sex characteristics
l) conception
m) clitoris
n) morals
o) testosterone

Changes and Responsibilities Chapter 2 35

Part B Write the word or words from the Word Bank that best complete each sentence. Not all of the words will be used.

WORD BANK	
adolescence	hypothalamus
moral	primary sex characteristics
erection	
nocturnal orgasms	menstruation
growth spurt	secondary sex characteristics
pituitary gland	ovulation

Puberty begins the stage of life known as _____. Many physical changes occur during this period of transition. A rapid increase in height and weight, known as the _____, often occurs. This is triggered by a small area in the brain called the_____. The _____—the male or female reproductive system—undergo tremendous changes. In females,_____ occurs as part of a regular menstrual cycle. In both males and females, _____may be experienced. These are normal and may simply be a sexual release for the body. Besides physical changes, this stage of life brings with it intellectual, psychological, social, and _____ development.

36 *Chapter 2 Changes and Responsibilities*

Chapter 2 Critical Thinking

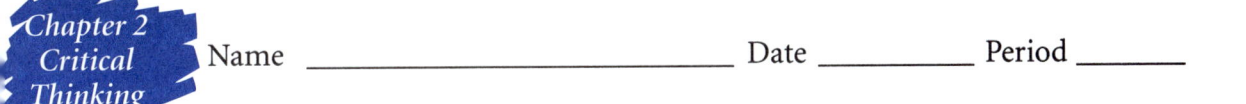

Name _____ Date _____ Period _____

Read the excerpts below from a magazine article. Then answer the questions that follow.

A Gift From Mother Nature

What sours wine, blights crops, kills bees, rusts bronze and drives dogs mad? Menstrual blood, claimed the Roman scholar Pliny the Elder. Sex during one's period was thought to cause the birth of monsters, and even as late as the 1970s, many women were told they suffered cramps because they hadn't accepted their "proper roles. . . ."

Menstrual superstitions . . . persisted even into [the twentieth] century. In the 1920s, the superstition circulated that a permanent wave [in the hair] wouldn't take during menstruation. Some musicians still blame broken violin strings on their own or their wives' periods. And one mother recalls that her own mother warned her: "Don't take a bath. And don't touch flowers, or they'll wither and die. . . ."

Slang expressions used to refer to menstruation suggest that even today the topic is still embarrassing and poorly understood. Expressions such as "on the rag" have come to mean cranky as well as menstruating. . . .

Even among themselves, many women do not mention menstruation directly. "The curse" refers to an old superstition, often stemming from [the biblical story of] Eve's expulsion from the Garden of Eden, and expressions such as "sick time of the month" and "being unwell" perpetuate the myth that menstruation is an illness rather than a normal bodily function. . . .

For centuries, menstruation has been shrouded in myth and mystery. Today, however, we have a better understanding of what this normal process is and how it works. And new research keeps adding useful knowledge. . . .

New Research

"There's a terrific blossoming of research on the menstrual cycle," says Alice Dan, Ph.D., a psychologist at the University of Illinois College of Nursing and president of the Society for Menstrual Cycle Research. "I credit it to the women's movement. There was demand to answer some of the questions asked by women."

Some of the most useful research has been on menstrual cramps. Scientists have recently found that substances known as prostaglandins (PG) cause uterine contractions and pain. Now medications with anti-PG action are being used to control cramps.

Menstruation is thus becoming less and less of a mystery and is now being recognized as a normal and healthy part of a woman's life. The true curse for far too long has been ignorance.

Consider the excerpt you have just read and remember what you have learned from the chapter as you answer the questions below.

1) How can ignorance about menstruation and other aspects of sexuality be harmful?

_____

_____

2) Define puberty. Explain why menstruation is an important aspect of puberty.

_____

_____

3) You have read that adolescence involves not only physical changes but also intellectual, social, and moral changes. What impact, other than physical changes, do you think menstruation has on a female adolescent?

_____

_____

4) How does health education help us to avoid misunderstanding sex and physical changes like menstruation? How does health education help us to avoid discrimination?

_____

_____

Chapter 3
Female and Male Roles

▶ What are female and male roles and how do they affect you?

▶ Where do we learn about gender roles?

▶ How might gender roles affect relationships and marriage?

▶ What is a healthy gender role?

Ms. Long had divided her health class into two groups—one boys, the other girls. Each group made up two lists. The first list was made up of "Masculine Traits and Characteristics," and the second list, "Feminine Traits and Characteristics." The two groups were in the process of verbally sharing these lists. The boys claimed that they were physically stronger. The girls said that they are more social, nurturing, and caring. The girls believed that boys are less emotional and more analytical and competitive. Voices became louder as the boys' group exclaimed that men are more intelligent, dominant, and creative and have higher self-esteem. The girls groaned and shouted that the boys were deceiving themselves. "We can twist you around our little fingers if we want," roared one girl.

At this point, Ms. Long interrupted. "Let's all calm down. Then we can see where each group came up with their ideas about gender."

Gender identity
The self-image of being either male or female

Gender role stereotyping
Confining expectations of how males or females should act

What are gender roles?

The announcement "It's a girl!" or "It's a boy!" when a baby is born is the start of learning what it means to be male or female. By two years of age, most children can label themselves as being a boy or a girl. These children have developed **gender identity**—the self-image of being either male or female. The process of learning more about gender identity continues on from there. A little girl will learn "how little girls act," while a boy will learn "how little boys behave." Each of them is learning gender roles, or sex roles—the behaviors and traits expected of them because of their gender.

As we learn about gender roles, it is easy to develop **gender role stereotyping**. These are confining expectations of how males or females should act. We all have gender role expectations of everyone we meet or know. Girls may be critical of another girl who seems too "tomboyish." Likewise, boys may be critical of a another boy who doesn't behave as expected.

What are some common gender role stereotypes?

For years researchers have studied gender role stereotypes—gender roles that match an overly simplified general or traditional pattern. They have found that while some stereotypes are truly misunderstandings, others are still questionable. Some beliefs about gender are:

- Girls are more social than boys.
- Boys are more intelligent.
- Girls are less assertive.
- Girls have lower self-esteem.
- Boys are more logical.
- Boys are more creative.
- Boys are more dominant.
- Girls are better at taking suggestions.

No conclusive differences have been found between boys and girls regarding these traits:

- intelligence
- agreeability
- creativity
- timidity
- assertiveness
- shyness
- dominance
- emotionality
- achievement
- competitiveness

On the other hand, true differences found between girls and boys include the following:

- Boys tend to be more aggressive than girls, especially toward other males.
- Girls are generally quieter and less boisterous than males.
- Boys are usually better than girls are at physical problem solving, such as the ability to finish three-dimensional puzzles.
- Girls tend to have better verbal skills after age 10 or 11 than boys do.
- Girls are more sensitive to touch than boys are.

Female and Male Roles Chapter 3

> **Sexual orientation**
> *A person's gender preference regarding sexual relations*

- In physical size and strength, boys tend to be bigger and stronger than girls.

How do gender roles affect you?

Playing roles is a part of life. For example, you behave like a student when you are in school and like a son or daughter when you are with your parents. Usually, you can be yourself within either of these roles. If you find that your real feelings are not being expressed in a role, then that role may be too confining and narrow. As a boy, you may feel that you always have to act strong, independent, and secure. However, inside you may feel hurt and lonely. In this case, playing the role can be destructive. Similarly, as a girl, you may believe that you must always appear attractive, feminine, and polite. Deep inside may be a part of you that wants to be different from this. If so, your current gender role may need to be changed or expanded.

A healthy gender role allows you to be yourself as you move from one situation to another. It can be fun to experiment with different roles by making small changes in your hairstyle, dress, hobbies, and other activities.

What about sexual orientation and gender?

Sexual orientation refers to a person's gender preference when it comes to sexual relations. A heterosexual person prefers a partner of the opposite gender. In contrast, a person with a homosexual preference is attracted to someone of the same gender. Homosexual males are called *gay*; homosexual females are called *lesbian*. A bisexual person may be attracted to people of both sexes.

It is not true that you can identify someone's sexual preference by the way the person dresses, talks, or acts. Labeling someone can be harmful and irresponsible. While many theories have been written regarding sexual preference, no one is sure why some people have a particular sexual orientation.

Where do we learn about gender roles?

We learn about gender roles from several sources, probably the most important being our parents. Many parents use pink blankets for girls or blue booties for boys. They give trucks to boys and dolls to girls. Such treatment helps formulate children's ideas of their masculinity and femininity. Generally, parents teach traditional roles, especially when it comes to fathers interacting with daughters and mothers with sons. If the same-gender parent is not involved in the child's life, the child may have a more difficult time learning his or her role.

Another way we learn about gender roles is from our peers. People in your age group strongly influence how you dress, what you say, how you act, and even how you feel. Females tend to imitate their female peers in certain ways, and males tend to imitate their male peers.

We also learn about gender roles from printed materials. The language, photographs, and illustrations in many books portray boys as action oriented and girls as more passive. Girls may be portrayed as followers rather than leaders.

Other sources for learning about gender roles include movies, song lyrics, music videos, television, magazines, newspapers, and books. As people grow older, they may take on some of the gender-role behavior seen in these sources.

How might gender roles affect relationships and marriage?

As people move from friendship to dating and from engagement to marriage, their gender roles may overshadow one another. People in relationships sometimes try to change their roles in the hope that their partner will like them better. People naturally want to behave in ways that please their partner.

Let's assume you are in a relationship in which you are playing a role. As the relationship continues, you learn more about what your partner likes. You start acting in more and more ways that satisfy your partner. Over time, you may become like an actor reading a script to make the other person happy. You stop being a real person. In some cases like this, people have played their roles so well that they become unsure of who they really are.

What gender role characteristics would you want a marriage partner to possess?

Some roles are limiting. They do not allow for relaxed responses or even surprises. Playing a role can also be boring for you and for your partner. Strictly playing roles will not build a healthy relationship.

While a division of labor may help some relationships run more smoothly, crossing boundaries can be healthy. Changes in our society and economy have dictated a need for changing roles. Women are needed in the workforce, and often, a family needs two incomes. Men share in parenting and household tasks more now than in generations past.

Individuals should get to know one another as people rather than as role players. Mutual respect and honesty are necessary for this to happen. Revealing things about oneself may show or even cause differences in values, goals, or needs. However, this kind of conflict can be healthy. By knowing these differences, the couple can choose to work with them or to find more appropriate partners.

Among couples today, you may find that:

- The man plays a large part in parenting of children.
- The woman earns as much as or more than her husband.
- The man is supportive, caring, and at times, passive.
- The woman is assertive and enjoys sexual relations as much as her partner.
- Both partners are comfortable with the woman taking more control.

Cathy© Cathy Guisewite, Reprinted with permission of Universal Press Syndicate. All rights reserved.

Chapter 3 Female and Male Roles

How do gender roles affect touch and communication?

Gender roles are a definite factor in how we communicate through touch. We know that touch is comforting and necessary for the health and well-being of infants. As we grow older, our comfort level with touch can affect the way we interact with people.

It is common for young children to touch their genitals for exploration and pleasure. How a parent reacts to this behavior can greatly influence how the child comes to view touch. Some parents tell their children that self-touching in private is all right. Other parents explain that this behavior is wrong and should not be done.

Masturbation—self-touching or self-stimulation of the genitals for sexual pleasure—is common among adolescents. Both boys and girls, or men and women, may masturbate. Throughout history, people have claimed that masturbation causes blindness, acne, or nervous disorders. Masturbation, however, cannot physically harm a person. Even today you may hear the myth that you are less of a person if you must resort to masturbation instead of "the real thing." This myth is not true.

Some people view masturbation as a way to relive sexual tension, while some feel no need to do it. Still others simply believe that masturbation is morally wrong.

During the first few years of school, children may want to hug or hold a friend as an expression of affection. Peer or parental influence can dictate what a certain child will do. Most often, friends at this age, especially boys, wrestle or scuffle.

A third-grade boy may pull a girl's hair or slap her on the shoulder to let her know he likes her. Of course, the girl probably will not see it that way.

In the junior high or middle school years, adolescents may get involved in different party games that involve touch. They learn quickly that touch feels good. The risk of getting caught may add to the excitement. Although this type of activity is usually played when adults are not around, the girls in the group frequently take on the parental role. It somehow becomes their responsibility to say, "That's enough," or "Don't go any further!" The boys and girls in this group have taken on the gender-role idea that boys are aggressive and girls must resist. Unfortunately, this type of situation pits boy against girl. If the boy wins, the girl loses.

Throughout the years, some people have developed certain "lines" to overcome their partner's resistance to having sex. One of the most unfortunate is the insincere "I love you." When spoken insincerely, this line is really a lie that reduces the value of love and sex for both genders.

Think about the behaviors of one boy and girl who act according to traditional gender roles. The boy does the touching while the girl allows herself to be touched. This may happen because they believe that the male is supposed to be the aggressor. Or it may happen because it frees the girl from some of her sexual responsibility. The girl in this situation learns she is desirable and has power to please or control the boy she wants.

The boy in this situation learns that touching a girl's concealed body is exciting. He believes that his fondling is just the "beginning," and he can barely wait to move on. While he may think that he is fondling, the girl may see him as grabbing or pulling. He may wonder why she is not just as turned on as he is. She may feel that he is thinking only of himself. She may let him continue just to keep him interested. This boy and girl, however, need to evaluate honestly why they are participating in this type of sexual activity. If they don't question their actions, they may split up and move into a new relationship in which the same thing could happen.

Many people come to think that touch is a means to an end. Numerous magazines and marriage manuals support the idea that men with the "right" touch will make a woman want

> **Androgyny**
> *The state of having what society defines as both masculine and feminine traits*

them. Readers are made to believe that hugs lead to kisses, which lead to fondling, and so on. In contrast, many girls feel that holding hands or kissing are pleasurable activities in their own right. They do not have to lead to any other kind of sexual touching. This kind of touch can be an expression of love, support, and caring and can be exciting.

In healthy marriages, sexuality involves emotional sharing through both touch and words. This kind of sexual expression is fulfilling because the partners feel secure, loved, and intimate as well as sexual. Troubled couples usually have to relearn how to touch this way, with an emphasis on communication rather than just on sex or individual pleasure.

What is the future of gender roles?

Gender roles will always be with us, but they are changing in the direction of **androgyny**. Androgyny is the state of having what society defines as both masculine and feminine traits. For instance, a man who is considered a "good provider" financially can also provide emotional comfort and nurture his partner. A woman can be both assertive and soft spoken.

What is a healthy gender role?

A healthy gender role is one based on honest personal needs and values. This allows greater freedom to express one's real self rather than simply creating an image. Therefore, a man can do things that some see as feminine and not have his masculinity questioned. He may sew or do dishes and still feel masculine. A healthy woman can have a career or drive big rigs and still feel feminine.

What do you think?

Draw a line down the middle of a sheet of paper. On the left, list the qualities you look for in a dating partner. On the right, list the qualities you want that desirable dating partner to see in you. Use as many adjectives as possible in both lists.

1) In what ways do your two lists differ?

2) What qualities on the left side of the page do you possess?

3) What qualities on the left side of the page would you like to possess?

4) How could you develop the qualities you would like to possess?

5) Compare your lists with those of your classmates.

How will you decide?

Explore the gender roles of your parents, one male peer, one female peer, and someone you admire. Write how you see their masculine, feminine, or androgynous traits. If appropriate, find out how these people view their own gender roles. How do typically masculine or feminine traits fit into their lives at home, at school, at work, and at play?

Choose some of the characteristics, behaviors, lifestyles, or activities related to gender roles that you admired the most in your exploration. Write these down and add to them to form a "new and improved" gender role for yourself.

Evaluate your new role by trying it. For two weeks, try to adopt or adapt one of your new gender-role ideas or behaviors into your own life. Monitor the effects this change has on yourself, your parents, and your friends.

Exploring the gender roles of others can help us decide whether we are developing a healthy role for ourselves. Decide whether you want to continue, modify, or abandon the new role that you've developed in this exercise.

Chapter 3 Review

Ideas to Remember

■ Gender roles are the behaviors and traits expected of us because we are male or female.

■ Gender identity is the self-image of being either male or female.

■ Gender-role stereotypes are confining expectations of how all men and women should act.

■ Men and women have more in common than they have differences.

■ Parents, school, peers, and popular culture are all sources for our gender-role beliefs.

■ You, your friends, and your relatives all can be negatively affected if your gender role is too restricting.

■ Androgyny is a state of having both masculine and feminine traits.

■ A healthy gender role is one based on honest personal needs and values.

Questions to Ask Yourself

1) What is the difference between gender role and gender identity?

2) What are five traits that males and females have in common?

3) Where did you learn most of the characteristics that make up your own gender role?

4) How does your gender role affect your relationships with others?

5) Would you say your gender role is a healthy one? Why or why not? If not, what needs to be changed?

Action for Health

CHECKING EXPECTATIONS

Work as a class to explore gender-role expectations. Imagine that you each have just become a parent. List the characteristics that you want your baby to have. Make one list of all these traits on the board. Decide as a group which traits are masculine, feminine, or androgynous.

Repeat the activity. This time, imagine you are a business owner listing the traits of an employee you are seeking. Decide which traits fall in each category as you did before.

Chapter 3 Review Sheet

Name _____ Date _____ Period _____

Choose the word or words from the Word Bank that best complete each sentence.

WORD BANK	
aggression	homosexual
androgyny	masturbation
gender identity	myth
gender stereotyping	parents
girls	sexual orientation
heterosexual	verbal skills

1) The possession of both masculine and feminine traits is called _____ .

2) One difference between boys and girls that has been documented is _____.

3) After age 10 or 11, girls are usually better than boys in _____ .

4) The existence of confining expectations about what a boy or girl should be is _____ .

5) The self-knowledge of being a boy or a girl is called _____ .

Female and Male Roles   Chapter 3

6) The idea that you can tell someone's sexual preference by the way he or she dresses, talks, or acts is a _____ .

7) A person's sexual preference in sexual relations is called _____ .

8) A person who prefers a partner of the opposite gender is _____ .

9) The most important influence and source of information about gender roles is _____ .

10) A person who prefers a partner of the same gender is _____ .

11) Self-touching of the genitals for sexual pleasure is called _____ .

12) In the junior high years, in group situations, _____ often assume the role of parent.

Chapter 3 Critical Thinking Name _____ Date _____ Period _____

The excerpt below is from a *Newsweek* article entitled "Guns and Dolls." Read it carefully. Then answer the questions that follow.

For 60 years, America's children have been raised on the handiwork of Fisher-Price, makers of the bright plastic cottages, school buses, stacking rings and little, smiley people that can be found scattered across the nation's living rooms. Children are a familiar sight at corporate headquarters in East Aurora, N.Y., where a nursery known as the Playlab is the company's on-site testing center. . . .

According to Kathleen Alfano, manager of the Child Research Department at Fisher-Price, kids will play with everything from train sets to miniature vacuum cleaners until the age of 3 or 4; after that they go straight for the stereotypes. And the toy business meets them more than halfway. "You see it in stores," says Alfano. "Toys for children 5 and up will be in either the girls' aisles or the boys' aisles. For girls it's jewelry, glitter, dolls and arts and crafts. For boys it's model kits, construction toys and action figures. . . . Sports toys, like basketballs, will be near the boys' end. . . ."

Even where no stereotypes are intended, the company has found that some parents will conjure them up. At a recent session for 3-year-olds in the Playlab, the most sought-after toy of the morning was the fire pumper, a push toy that squirts real water. "It's for both boys and girls, but parents are buying it for boys," says Alfano. Similarly, "Fun with Food," a line of kitchen toys including child-size stove, sink, toaster oven and groceries, was a Playlab hit; boys lingered over the stove even longer than girls. "Mothers are buying it for their daughters," says Alfano.

Children tend to cross gender boundaries more freely at the Playlab than they do elsewhere, Alfano has noticed. "When 7-year-olds were testing the nail polish, we left it out after the girls were finished and the boys came and played with it," she says. "They spent the longest time painting their nails and drying them. This is a safe environment. It's not the same as the outside world."

Consider the excerpt you have just read and remember what you have learned from the chapter as you answer the questions below.

1) Define "gender stereotyping" and give an example of it from the passage on the previous page.

_____
_____
_____

2) What evidence does the passage give to support the notion that environment has an effect on how boys and girls develop gender roles?

_____
_____
_____

3) In the Playlab, boys played with toys that are traditionally for girls, such as nail polish and a play kitchen. Why do you think that outside of the Playlab, boys and girls usually play with toys traditionally associated with their gender? Use information from the text to support your answer.

_____
_____
_____

4) Think of the toys you played with as a child. Did these toys encourage gender stereotyping? Give examples to support your answer.

_____
_____
_____

Chapter 4
Becoming Sexually Responsible

- What is sexually responsible behavior?
- What is high-risk behavior?
- What is the pathway to sexual responsibility?
- What steps are involved in making decisions?
- How can you set limits and handle the pressure to have sex?
- What are your major responsibilities?

Tom and Mary were sitting in their family room going through their high school yearbook. "Remember Cindy?" Mary asked. "She dropped out of school in her last year because she was pregnant. For a long time I didn't know why she was so popular with the boys, but her rep and let's-party attitude caught up with her. Look! Here's Eddy."

"Yeah," laughed Tom. "None of the guys wanted to shower with him after PE because we always heard he had the 'clap' or some other kind of disease. Unfortunately, a lot of girls found that out the hard way."

"Here we are at the prom!" Tom pointed out. "I knew then that I wanted to marry you. But I'm glad we waited until after college."

"Me, too. I still have grad school ahead of me—but you have a degree, a good job, and me!" Mary exclaimed.

"I think we do okay making all the important decisions. That's why I think we'll know when it's right to start a family," Tom said confidently. "Whatever happened to Suzanne?"

"She dated Gary for about eight months until he raped her. She was so devastated that she transferred out during senior year," Mary explained.

"I wonder why some of us handled the pressures of sex pretty well and others seemed to have so much difficulty?" said Tom. "We wanted to have sex as much as anyone else, but it paid for us to wait."

> **Sexual responsibility**
> *The ability to speak and act honestly and directly, and to show respect and care in a relationship*
>
> **High-risk behavior**
> *Actions or activities that place a person in greater than average danger*

What is sexual responsibility?

Sexual responsibility means being honest and direct in words and actions. Being sexually responsible in a relationship involves showing respect for the other person and caring about how the person feels. Both people work to make sure that neither takes advantage of the other. They don't avoid sexual responsibility with excuses like, "I didn't mean to go so far" or "I was drunk and didn't know what I was doing." They don't allow themselves to be pressured into unwanted behaviors, and they don't put such pressure on others.

Sexually responsible people are aware of how powerful sexual emotions can be. They realize how complicated situations can become if they ignore sexual responsibility. They know that responsible sexual decisions prevent problems. Cindy, Eddy, and Gary did not use sexually responsible behavior.

Paying attention to your own behavior and feelings can help you know whether you are mature and sexually responsible. Do you say or do things that go against your values or bring harm to yourself or others? If you do, then you probably are not being sexually responsible. It's like eating a triple-dip ice cream cone when you want to maintain your weight or smoking behind closed doors when you said you would quit. You may get some enjoyment out of that brief moment. Unhappy feelings or guilt may well overcome that slight pleasure, however, because your behavior was untrue to your goals or values.

What is high-risk behavior?

High-risk behavior is actions or activities that place a person in greater than average danger. For example, you choose to ride in a pickup truck without wearing your seat belt. This choice puts you in greater danger of being injured in an accident than if you wore a seat belt. You are at even greater risk if you elect to ride in the open back of the pickup truck. Similarly, certain sexual behaviors place you at great risk for:

- getting a sexually transmitted disease such as AIDS
- becoming pregnant or getting someone pregnant
- finding yourself in potentially embarrassing situations that go against your values

A sexually responsible, self-respecting individual avoids such high-risk behavior. People can reduce their risks by changing their behavior. On some occasions, however, people become victims, not because of their own behavior but because of someone else's behavior.

What is the pathway to sexual responsibility?

The capacity to act in a sexually responsible fashion requires an understanding of yourself as a sexual being. You need to know the power of sexual feelings and the possible results of sexual actions.

Achieving sexual responsibility takes time as you progress through adolescence. You go through a process of continual change. As you learn to handle simple issues and responsibilities, you become better prepared to deal with complex decisions. This process will happen naturally if you give yourself time. Each day you have new experiences that teach you something about yourself. Like a flower bud, you slowly unfold to become known to yourself and to others.

How does an opening flower bud symbolize the pathway to sexual responsibility?

The knowledge you gain from this process can be helpful as you continue to mature and change. For instance, you know that Tom and Mary were very much in love in high school. However, they knew having sex and getting pregnant were not options for them. This sexual self-knowledge and restraint played a part in helping them achieve their goals. They graduated from college, and Tom got a good job. Their path continues together as Mary prepares for graduate school. They are thinking about having children. Choosing to abstain from intercourse was an important step in their maturing. Although Tom and Mary were teenagers with sexual desires, they had matured enough to set limits on their behavior. Then they were able to reach more difficult goals they had set for later in life.

Some of the processes of maturing sexually are:

- development of a positive, realistic body image
- the ability to overcome fears that produce needless guilt or shame (For example, you may choose to ignore the myth that masturbation makes you crazy.)
- an understanding of sexual activity that is appropriate to your current age and lifestyle (An example is realizing you are not in a position to parent a child right now.)
- a gradual loosening of ties to parents, siblings, or other family members
- clarity about gender identity or sexual orientation
- the ability to distinguish between positive and negative aspects of sexuality (For example, X-rated movies that show people as sexual objects are negative stimuli, and taking advantage of a girl who has been drinking is wrong.)
- the skill within a relationship to bring love and sexuality together in a meaningful way
- the ability to act responsibly toward yourself, your partner, and society

The degree and direction of maturing varies from person to person. Your biological makeup, your experiences, and your environment will dictate how you mature. As you mature, you will gradually learn that truly feeling good about yourself must start inside you. It will not be dependent on what your friends or peers think, do, or say.

How long it takes to move through this pathway of maturation is different for each person. For most American adolescents, the time frame for moving through the pathway is from puberty, or around age 12, to the early or midtwenties. A great deal happens during these years.

There are major growth changes physically, intellectually, socially, and morally. Also, there may be many sexually related experiences, from parties to dating to perhaps a loving, meaningful relationship with one partner. Another reason why people mature gradually is because it takes time to learn accurate facts about sexuality. Keep in mind also that while most maturation usually takes place during adolescence, maturing as a human being is a never-ending process.

What can you expect to learn along this pathway?

To help yourself mature naturally and positively, you must learn something about yourself and your sexual self each step of the way. Following are some examples of steps along your pathway, what you should learn from them, and some expectations you should have.

Early dating, games, friends During early dating, it's important to learn to talk with your partner as a friend. Go slow with the relationship by setting limits, first on your own, and then together. "Playing games" to manipulate another person is damaging and immature. Learn to see members of the other gender as potential friends rather than as objects, conquests, or enemies.

Touching and your own privacy In many dating relationships, you must learn how much private space you are willing to share with your partner. Early on in a relationship, you may be comfortable holding hands with your partner. As the relationship progresses, new questions will arise. How close should we be during a slow dance? Is it okay for us to kiss now? What if a tongue is used? What about fondling?

Setting limits It is important to set limits in advance on how far you will go in a sexual relationship. Your own limits will help prevent you from doing things you might regret.

Comparing body images Rating your own body, especially during the changes of puberty, is a natural thing to do. But it is important to be careful. For instance, if a girl looks at a carefully photographed and electronically changed picture of a model, she may feel unattractive. A boy may see himself as lacking if he reads stories in magazines that tell of sexual skill. In fact, measuring oneself or one's partner against unrealistic standards may be a sign of immaturity.

Giving and receiving love Much of what you learn about giving and receiving love comes from your family. This can be a positive or a negative influence, depending on your family. When you move into an intimate relationship, you must treat your partner the way you want to be treated. It is important to see your partner as a respected equal with whom you can share your innermost feelings.

Compromising You may feel that the limits you set are challenged by your partner, by what other couples are doing, or by things you learn about yourself. Many individuals fail to discuss limits with their partner. They just "let it happen." In a sexually responsible relationship, the couple talks together to arrive at a mutual decision about limits. For instance, moving from petting to intercourse is a huge step. Taking this step means the relationship moves to a level for which each person should be ready and suited. Important possible results such as disease or conception as well as contraception must be considered. Most sexually active adolescents do not yet have the skills and experience necessary for making responsible decisions regarding intercourse and intimacy. These kinds of decisions are sometimes difficult even for adults, so it is important to discuss them together.

What steps are involved in making decisions?

While moving along the pathway to sexual responsibility, you can improve your decision making by following these steps.

1) Clearly describe the problem for yourself. Writing the problem down on paper helps you to focus on the most important part of it.

2) List on paper all possible solutions to the problem.

3) Think about the outcomes of each solution. Ask yourself: How will each solution affect my goals and values? Is the solution realistic? Will it be positive for my emotional and physical health?

4) Act on the best possible solution. Keep your eyes wide open. You can still change your mind. Also, you may learn something important, even from a negative outcome.

5) Evaluate the success of your solution. If there are new or remaining problems, try something else. You also might ask others for help.

As you mature, your decision-making skills will improve. The sexually responsible person thinks ahead, preparing for decisions in advance. Developing your decision-making skills will help you in many other areas of your life.

How can you set limits and handle the pressure to have sex?

If the world was perfect, no one would take part in a certain sexual behavior unless it was the right thing for that person. Everyone would have the experience, self-knowledge, and maturity to have sex only in a comfortable, meaningful relationship. Behavior would match a person's level of maturity. However, we do not live in that kind of world.

You may find that in your world, the values of parents are ignored or rejected. Cultural or religious values may seem unimportant. Many young people have the attitude that "everyone's doing it." Like many adolescents, you may be expected to decide between your own original limits and the standards your friends set.

In order to set limits, you may need to consider a few more questions. For example, did you know that most teenage boys leave a relationship if the girl gets pregnant? Are you aware that most teen marriages fail?

Do you recognize that some boys who act mature are merely using "lines" (lies) to get sex? Is a girl simply deceiving herself

THE FAR SIDE By GARY LARSON

© 1986 Universal Press Syndicate

"So, what d'ya say?.. Maybe we could go back to my place... Have a few drinks... a little alfalfa... Maybe we could show each other our brands... Ha ha ha ha ha"

Cattle hustler.

by saying "we made love" when she and the other person simply had sex? If a boy believes "nice girls shouldn't have sex" yet tries to convince girls to do so, isn't this is a put-down of all girls? If people think they must have a partner no matter what, might this mean that they are simply unhappy with themselves?

Did you know that the number of cases of genital warts (HPV) is increasing among young people? Do you know that some strains of HPV can cause cancer? Other STDs are becoming more common among teenagers than in the general population. What do you know about other sexually transmitted diseases and how to prevent them? What about contraception?

It is important to be able to answer these and many other questions.

What are your personal limits?

In addition to having basic knowledge about sexuality, you must determine what behaviors are morally right for you. Perhaps the values your parents or religion have given you might work, at least for the time being. Although you may question these values, they may allow you time to mature and to develop your own personalized values and beliefs.

One way to keep the limits you set is to avoid situations that might challenge them. Here are some ideas to help you keep your limits:

- Double date if you feel you need to avoid being alone with your partner.
- Go on a group date and do such activities as bicycling, bowling, or skating.
- Go to a dating partner's home only when one or both parents are present.
- Use your parents' time limits as a reason to leave early if needed.

- Call parents for a ride home if necessary.
- Listen for information on how your dating partner has treated others.

When you have set your limits, let your partner know what they are. If your partner ignores your wishes, that tells you something you need to know about the person. Above all else, believe two very important words—"I count." Repeat this silently to yourself right now several times. If you really believe this phrase, you will make sure that your wishes are respected.

Finally, acting on your limits and asserting yourself require self-discipline. Think of one or two areas in which you have self-discipline in your life. Think of one or two areas in which you would like to have more self-discipline. As is true for many things, self-discipline takes practice.

What are your major responsibilities?

A simple answer to that question is that you need to be responsible for yourself and all of your actions. For sexual responsibility in particular, you need to:

- Gain knowledge about sexuality and sexual health.
- Choose a value system that makes sexuality a positive force in your life.
- Talk with parents, teachers, and counselors when necessary.
- Set your own limits and make them known to partners.
- Stick up for yourself, saying and believing "I count!"
- Report incest, rape, sexual abuse, and the spread of sexually transmitted diseases to appropriate authorities.
- Living up to these responsibilities may be difficult, but it will certainly be worthwhile. Following them will strengthen the "real" you in an honest, direct way.

What do you think?

Read the following passage from *Zorba the Greek,* by Nikos Kazantzakis.

I remember one morning when I discovered a cocoon in the bark of a tree, just as the butterfly was making a hole in its case and preparing to come out. I waited a while, but . . . I was impatient. I bent over it and breathed on it to warm it. I warmed it as quickly as I could and the miracle began to happen before my eyes, faster than life. The case opened, the butterfly started slowly crawling out and I shall never forget my horror when I saw how its wings were folded back and crumpled; the wretched butterfly tried with its whole trembling body to unfold them. Bending over it, I tried to help it with my breath. In vain it needed to be hatched out patiently and the unfolding of the wings would be a gradual process in the sun. Now it was too late. My breath had forced the butterfly to appear . . . before its time. It struggled desperately and . . . died in the palm of my hand.

1) What is the message of this passage?
2) How does this message relate to sexual maturation?
3) How does this passage relate to abstinence?
4) How might this message apply to you or to your relationship with a dating partner if you have one?
5) Where are you along the pathway to full maturity and sexual responsibility? Relate your answer to the passage.

How will you decide?

Explore the things that some people do or say to convince others to take part in sexual activity. Write out a list of such "lines" or "moves." Brainstorm with the class a list of responses or behaviors to persuade the person to "back off."

Choose one or two responses you feel are strongest. Which are more likely than others to work? Keep in mind that harsh put-down responses may cause anger or even abuse.

Ask for a friend or parent's opinions. Try using one of the responses if an appropriate situation arises.

Chapter 4 Review

Ideas to Remember

■ Sexually responsible people are able to choose between right and wrong. They are honest and direct in words and actions.

■ Sexually responsible people show respect by letting their partners know how they feel and that they will not take advantage of them.

■ If you say or do things that harm yourself or others, then you are probably not yet capable of sexual responsibility.

■ Sexual responsibility is often not fully acquired until the early to midtwenties.

■ Setting and keeping limits through self-knowledge, discipline, and strong decision-making skills demonstrates sexual responsibility.

■ High-risk behavior places you in greater than average danger and indicates you are still maturing.

Questions to Ask Yourself

1) What is the meaning of sexual responsibility?

2) How can you tell if a person is being sexually responsible?

3) What are your limits for sexual behavior?

4) What are some high-risk behaviors in regard to your sexuality?

5) Why is abstinence the only way to eliminate sexual risks?

Action for Health

REASONS FOR LIMITS

Write a list of reasons teenagers might use for having sexual intercourse. Then add a list of possible consequences to having sexual intercourse. Third, write a list of reasons for teenagers to abstain from having sexual intercourse.

Discuss all the lists and decide:

- Which reasons have a greater impact on a teenager's present life?
- Which reasons are most likely to change a teenager's future?
- Which reasons do movies, television, and printed materials most emphasize?

Through a class vote, eliminate all "weak" reasons for having sexual intercourse. Finally, discuss how abstinence fits into the pathway to sexual responsibility with responsible decision making.

DEFINING YOUR LIMITS

One definition of the term *abstinence* is "self-denial with regard to hunger, pleasure, or craving." For some teenagers, this definition might be accurate. But for others, abstinence is a healthy, desirable choice that does not require great effort. These people feel that making a choice for their well-being is easy. Rather than self-denial, abstinence for them is self-fulfilling.

Look up the term *abstinence* in several dictionaries. Write the definitions on index cards. As a group, study all the definitions. Look back at the reasons to abstain from having sexual intercourse that you listed in the Action for Health activity above. Finally, come up with a new positive term or phrase that you could use in place of the term *abstinence*.

Chapter 4 Review Sheet

Name _____ Date _____ Period _____

Circle the letter of the answer that correctly completes each sentence.

1) An example of sexually responsible behavior is:
   a) understanding how sexual activity brings physical pleasure
   b) learning how to go on dates
   c) considering the consequences of sexual activity
   d) not engaging in sexual activity

2) Which of the following is not an example of sexual maturity?
   a) absence of confusion about gender identity
   b) development of a positive body image
   c) feeling guilt about masturbation
   d) loosening ties to parents and other family members

3) Engaging in numerous sexual encounters is an example of:
   a) high-risk behavior
   b) sexual responsibility
   c) sexual maturity
   d) giving and receiving

4) The largest part of sexual maturation occurs:
   a) as a child
   b) as an adult
   c) in later life
   d) in adolescence

5) Limits on sexual relations while dating should be set:
   a) as the relationship progresses
   b) in advance
   c) after a few dates
   d) when both partners agree to set them

Becoming Sexually Responsible   Chapter 4   **69**

6) Comparing body images is:
   a) unnatural
   b) normal
   c) a sign of immaturity
   d) unrealistic

7) A good example of negotiating is:
   a) setting limits mutually
   b) letting fate take its course
   c) moving from petting to intercourse
   d) discussing sexual activity with friends

8) Which of the following is not a part of decision making?
   a) considering the consequences
   b) brainstorming possible solutions
   c) rejecting all outside suggestions
   d) evaluating what happened

9) Which of the following statements about teenage sex is true?
   a) Pregnancies among teens are increasing.
   b) The cases of STDs among teenagers is finally dropping.
   c) When a teenage girl becomes pregnant, the father usually stays in the relationship.
   d) Most teen marriages succeed.

10) Which of the following is not a good way to enforce your limits on sexual activity:
   a) double dating
   b) spending time alone with your partner
   c) group dating
   d) obey time limits

Chapter 4 Critical Thinking

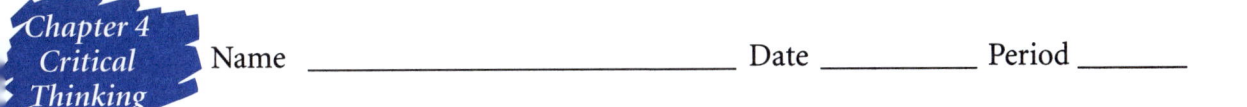

Name _____ Date _____ Period _____

Do you think we can possibly learn something about human behavior by studying animals? Read the excerpt below from an article in the *Los Angeles Times*. Then answer the questions that follow.

> Like teenagers who have to pass up a hot date on Saturday night because their parents make them stay home and baby-sit, some birds who want to mate must overcome the opposition of their elders.
>
> Adult male white-fronted bee-eaters, common birds on the savannas of eastern and central Africa, actively disrupt the mating attempts of their male offspring and other males in order to recruit the adolescents into helping with the care of their younger siblings, Cornell University researchers reported . . . in the British journal *Nature*.
>
> In a clear demonstration of instincts honed by evolution, the older birds chase their sons away from potential mates, prevent them from giving food gifts, and block access to nesting sites.
>
> The tactic works. Young males were twice as likely to abandon their mating quest and return to the family nest when harassed as when they were left alone.
>
> Why don't the sons put up more of a fight? Because by helping to ensure the survival of its siblings, say the scientists, the son increases the likelihood of its own genes being passed on to subsequent generations, even if it does not pass them on itself. . . .
>
> White-fronted bee-eaters, colorful birds about the size of a Western bluebird, live in colonies of as many as 200 birds. The young remain at their parental nest until they mate at one or two years of age. When they mate, males build new nests in the same community as their parents, but females leave to join the community of their new mate. About half of all nesting groups include young non-breeders who help provide food and protection for their younger siblings.
>
> Emlen and Peter H. Wrege of Cornell studied two populations of individually marked, white-fronted bee-eaters for five breeding seasons at Lake Nakuru National Park in Kenya and documented the generational conflict.
>
> Virtually all the harassers—91%—were male, about 65% were the fathers or grandfathers of the harassed birds and three-quarters were older than the

individuals they harassed. In addition to chasing the younger males and interfering with food gifts, the older birds would make repeated visits to the pre-nesting site of the younger birds to disrupt the mating.

About one-third of the time, the harassed bird would become frustrated and return to the home nest.

Consider the excerpt you have just read and remember what you have learned from the chapter as you answer the questions below.

1) Explain how the parent birds help their children to be sexually responsible.

___

2) What purpose does the parents' harassment serve for the bird community?

___

3) What purpose do you think being sexually responsible serves for human beings? Use your text to help you answer.

___

4) List three things you need to do to become sexually responsible.

___

Chapter 5
Communication and Relationships

- Why is communication important?
- Why is it so difficult to communicate about sexuality?
- What is the difference between a friendship and an intimate relationship?
- How do you assert yourself?

"For this concert," Teisha said, "I feel that I should be able to stay out until 1:00 A.M."

Her father commented, "You might be on your best driving behavior, but what about other kids who might be driving drunk or stoned? I wouldn't want anything to happen to you."

"You could park farther away and take a shuttle to the coliseum," Teisha's mother suggested. "There wouldn't be so much congestion or hassle."

"I can live with that," Teisha smiled.

Across town, Venessa was talking to her boyfriend, Tony. "I think we should date other people," she said.

"Come on," Tony said somewhat loudly. "You always need to have things your way!"

"You make me want to scream," interrupted Venessa.

Tony was settling into the fight. "Six months ago you wouldn't have said that. In fact, you were getting your own way back then, too," he shouted. "Maybe it's time to just forget about us."

Self-disclosure
Revealing your real self and honest feelings to other people

Why is communication so important?

Good communication is important—and extremely difficult—in any relationship. This is especially true if the topic is sexuality. Many families have excellent communication except when it comes to talking about sexual issues. Many couples who don't communicate well find that their relationship fails. Even though communicating well cannot solve every problem, it is essential for maintaining any healthy, lasting relationship.

Real communication involves **self-disclosure**. That means letting others know your real self and how you honestly feel. It means carrying on an open and honest dialogue with others that is based on love and trust. In a loving relationship, you want to know who the other person really is. You also want to let that person know you. Letting someone know who you really are can be frightening.

How easily do you self-disclose?

74 *Chapter 5 Communication and Relationships*

Parents who want their children to learn real communication allow them an early emotional freedom. These children are encouraged to show a full range of emotions—love, hate, anger, joy, sadness, fear, happiness, and so on.

Why is communicating about sexuality difficult?

In many homes, children are not encouraged to show much emotion. They may also be taught that sex is not a topic for discussion in the home. These children are not able to talk or learn about sexual topics. Parents may have their own reasons for this behavior, which are discussed later in the chapter.

Both children and adults may be unsure of the words to use when talking about sex. Many couples are even too embarrassed to talk about sexuality together, thinking the topic is too personal. Yet, these same people share their body with each other and even engage in high-risk behaviors.

Adolescents may have difficulty communicating with their parents because they are learning to be independent. They are trying to set new boundaries for themselves. If you are in this situation, remember that working with your parents to set your boundaries is possible. They know you well, they love you, and they can help. You can ask a parent to keep your conversations private from other family members. Seeking the advice of a parent can be like talking with an expert—something many adults do often.

How well do you express love, hate, anger, joy, sadness, fear, and happiness?

How difficult is communication about sex for you?

How can you communicate with your parents?

There are many reasons why some parents and children don't talk about sex. Some parents are uncomfortable with the topic and deal with it the same way their parents did. Other parents simply don't know how to say what they want to say. Parents may fear they will not know the correct answer to their children's questions. A few parents believe that discussing sexuality will make their children too curious about sex. It helps to try to understand why your parents may feel the way they do.

Most adolescents see their parents as out of step with today's teenage world. This might be true for many parents. Look at the items below and mentally check off those that you think apply to your parents.

- too strict
- don't care
- don't know anything
- old fashioned
- smothering
- frightened by the topic
- too busy
- too prying
- don't know me
- sarcastic
- don't understand the world today
- don't know what to say

Teens can take some responsibility for initiating and maintaining communication with their parents.

If you are already comfortable talking with your parents, you may simply need to make a transition into addressing sexual topics. In this case, you can let your parents know that:

1) You are glad you have "askable" parents.
2) If they do not have an answer to a question, you can look it up together.

3) They do not need to worry about your behavior just because you ask a question about a particular subject, such as contraception.

4) When you question values, it does not always mean you are rejecting them.

5) If you don't understand something, you will ask them to explain.

6) You know that information is not the same as permission.

7) You would like them to respect your opinions.

8) You may want them to establish some rules you can use as reasons for not doing something that peers suggest or for making it easier to say no.

9) You hope that both you and your parents will be forgiving, open, and supportive.

10) You are grateful for input, but you will need to make some decisions on your own as a part of growing up.

If you decide to approach a parent, choose the most helpful statements above. Practice starting a conversation. This kind of mental practice may make the real encounter much easier.

COMMON COURTESIES

- Make a point to spend a few hours a week with your parents.
- Every once in a while, ask your parents to tell you about their day.
- Occasionally ask one or both parents for their advice on something not too crucial or alarming. Try to follow their suggestions to show that you respect their judgment.
- Experiment by telling your parents the truth about issues, events, or feelings that you might normally have difficulty sharing with them. Start by letting them know you are taking a chance that they might become upset.
- Clean up your room or do other helpful chores at unexpected times.
- Praise your parents for the things you think they do well.

Mutuality
Having the same feelings for each other

Reciprocity
Mutual sharing or exchange of information between two people

If you do not feel that you have a good relationship with your parents, start by being polite and treating them like "nice people" for a while. After perhaps a month has gone by, let them know that you would like to (l) talk with them and (2) have them really listen. You can begin by discussing a topic that is not too threatening, such as your school or their work. To help you on your way to good communication at home, read the list of common courtesies in the box on the previous page. Your parents will appreciate these courtesies. These strategies may not always work for you and your parents, but they may bring rewards in other ways.

What role does self-disclosure play in friendship?

A friend is a person with whom you can talk and share your feelings. A friend can help you feel good about yourself. Friends respect one another, are good listeners, tell you the truth, and keep your secrets. All of this trust and respect occurs through the process of self-disclosure. Here are self-disclosure steps involved in becoming a friend.

Step 1: Choosing Choose a person you think you would like to have as a friend. You may notice that the person is attractive or shows some of the same interests that you have. As you initiate conversation, perhaps you will share just your name, where you live, and your year in school. Sometimes, the person you choose will not be interested in forming a friendship, which takes you to the next step.

Step 2: Mutuality This necessary step is simply that the person you choose begins to share information with you. You have reached **mutuality** when you have the same feelings for each other.

It is all right if someone doesn't choose you back. Another person could avoid forming a friendship for many unknown reasons. Don't feel put down. Many other wonderful people will want to know you.

Step 3: Equal Sharing When a true friendship evolves, your friend will share an equal amount of information with you. You have **reciprocity** when you mutually share or exchange information about yourselves. It may take time for you to feel the reciprocity.

Step 4: Trust As you and your friend share more of your feelings, you somehow know that the person will not repeat your feelings to anyone else. By sharing information about yourself, you expose yourself emotionally; however, a true and trusted friend will not take advantage of this.

What is the difference between a friendship and an intimate relationship?

In an intimate relationship, people share even more meaningful information than friends do. An intimate relationship is not necessarily a sexual relationship. An intimate relationship may start out the same way a friendship does, but it comes to involve deep self-disclosure. These partners become even more exposed emotionally to one another than friends do. As a result, they develop true intimacy. They know that their intimate partner can help them or truly hurt them, because they know each other so well. Developing a truly intimate relationship can be frightening. However, like a good friendship, it can last forever if both people work at it.

Generally, happily married couples have a lot of self-disclosure. They share almost all of their feelings with one another. Nonetheless, a part of each person usually always remains private—an inner core that belongs only to that individual.

In addition to sharing their deep feelings, intimate couples may also share their body through touch. This type of self-disclosure adds even greater risk. Some common fears are "What if she doesn't like my body?" or "What if he doesn't like the way I touch him?" These fears will disappear over time.

Unlike a dating situation, marriage partners do not just walk away from one another if difficulty arises. They try to work out their differences through even more self-disclosure. As you can see, being truly intimate has much more depth and meaning than just having sexual intercourse.

In this cartoon, Peter has just told his girlfriend, Denise, that he thinks they should date other people. In what ways are they communicating their feelings to each other?

Assertiveness
The ability to stand up for your own rights without violating another person's rights

Aggressiveness
Standing up for your own rights without regard for others' rights

Nonassertive
Giving up your rights so that other people can do whatever they want

What are the basics for good communication?

Good communication requires two components: an effective speaker and an active listener. The speaker's role is to deliver a clear message. The active listener's task is to make sure that the message heard was the message that the speaker meant. The active listener also gives feedback, as appropriate, through body language or reflecting back what the speaker said.

In some relationships, one person functions as the speaker most of the time, while the other functions as the listener. This habit should be broken, because the roles of speaker and listener need to be shared. The boxed lists describe some do's and don'ts for good communication. The items on these lists are especially important when trying to settle a disagreement.

How do you assert yourself?

Representing yourself in any relationship takes courage, self-disclosure, and **assertiveness.** Assertiveness means being able to stand up for your rights without violating another person's rights. By comparison, **aggressiveness** means standing up for your own rights without concern for someone else's rights. **Nonassertive** behavior implies that you are giving up your rights so that other people can do whatever they want.

For example, Jessica and Justin are at the end of their first date. As they stand at the door to her house, Justin leans over slowly to kiss her goodnight. If she responds aggressively, Jessica may shout, "Hey! What are you trying to do, you jerk?" She would be protecting her right not to be touched. But she would not be concerned about Justin's right to make a mistake and be treated courteously. A nonassertive Jessica would let Justin kiss her when she doesn't want that. She might even refuse to date him again without letting him know why. An assertive response from Jessica might involve a gentle push away from Justin while adding, "I'm really flattered, but I better go in now." In that way, she stands up for her rights and feelings but also respects Justin's.

Do's

I statement
A way to share feelings about something that you want to resolve without putting the other person down

1) *Use I statements to express yourself.*
   I statements are a way to share feelings about something that you want to resolve without putting the other person down. Many times people use you statements such as, "You always wait so long to call me. You are so selfish." This type of statement can cause the other person to become defensive. An I statement such as "I feel hurt when I don't hear from you" avoids defensiveness.

2) *Be specific.* Try to state your feelings as specifically as possible, especially when making a request for a behavior change. If you are not specific, the other person may misunderstand. For example, rather than tell a parent, "I'll be home early tonight," say, "I'll be home by 10:30 tonight." This leaves no room for varying interpretations. By the same token, you can ask for specific messages from your parents.

3) *Show respect, warmth, and caring.*

4) *Be honest.*

5) *Get involved as a listener.* Nod your head to show you understand and avoid interrupting. Ask questions when you are unclear about something.

6) *Maintain eye contact*—it's vital for good communication.

7) *Give feedback.* In your own words, restate what you think you heard the speaker say. Feedback like this lets both of you know that the message was clearly understood. Let your friend know you are glad he or she cares enough about your relationship to raise the issue.

8) *Consider compromise.* Your viewpoint counts, and so does your friend's. It is important to show respect for each other's viewpoint even if you may disagree.

Don'ts

1) ***Don't be sarcastic.*** Making smart comments or mean-spirited jokes causes your partner or friend to feel hurt or angry. Such behavior can close down communication.

2) ***Don't assume you know what the other person is thinking.*** Assuming you know what your partner is thinking can be destructive. If you want or need to know what the person is thinking, it is your responsibility to ask.

3) ***Don't use yes-no questions,*** because those answers won't tell you much. Ask open-ended questions. For example, instead of saying, "Did you like the party?" you might try, "What went through your mind at the party last night?" This method will be especially helpful if your partner is having difficulty opening up.

4) ***Don't play games or manipulate.*** Playing games tells your partner or friend that you really do not respect or care for him or her.

5) ***Don't use labels or name-calling.*** This is especially necessary to keep in mind during moments of anger. Labels or names can be hurtful.

6) ***Don't act uninterested or be impatient.***

7) ***Don't bring up old issues.*** Stick to the here and now. If old issues are still upsetting you, bring them up separately at a different time.

8) ***Don't overload your partner with a long list of complaints.*** Deal with them one at a time.

> **STEPS TO SAYING NO**
>
> **1)** Choose behaviors that are acceptable to you.
>
> **2)** Let the other person know that you appreciate the invitation: "I'm glad that you are attracted to me, but it's not right for me." Your response should be sincere.
>
> **3)** Say no in a clear manner: "I'd rather not kiss/make love/go dancing/get involved," or whatever.
>
> **4)** Provide an alternative to the activity you are saying no to: "I'd just like to hug for now." Use this step only if providing an alternative seems appropriate.

What is a good way to say no?

Saying no to unwanted behaviors, including sexual behaviors, can be especially difficult if you are caught off guard. To make saying no go more smoothly, many sexually responsible people develop a plan. The four-step plan above can be used in such a situation. Step 1 should occur well ahead of any sexual behavior. The next three steps should be used at the time the invitation to become sexually involved is made. Remember that an invitation will probably be made nonverbally through a touch, caress, or kiss.

This four-step plan is a positive way to say no. You have thanked the person for his or her interest while not following through with an unwanted behavior. Offering an alternative can change the subject and keep the other person from feeling rejected. It lets the other person know you are still interested in him or her but not in the unwanted behavior.

A key element in using assertive behavior is sending clear messages. A mixed message can leave the other person wondering what you really want. For example, it can be confusing and frustrating to a boy when a girl says she doesn't want to be kissed yet at the same time she seems to be flirting.

How should you handle criticism?

Being criticized by someone you care for can be upsetting. However, with honest self-disclosure, both positive and negative feedback may be expected. The way we respond to criticism can either help resolve the problem or bring a quick end to further communication. If the relationship is important, you both will work through the complaint to arrive at a fair solution.

One of the first things you can do when you are criticized is to pause and think about the complaint. Avoid becoming immediately defensive and blurting out something just to get even: "Oh, yeah? Well, what about you? You always . . ." Understand that the other person has risked making you angry to let you know that he or she is upset about something. Or perhaps the person wants to share something that could help you learn about yourself. If the person did not care about the relationship, he or she would not have taken this kind of risk.

After you have paused, repeat back what was said in your own words. This will eliminate any possible misunderstanding. If necessary, let the person clarify what he or she said. Ask for more information as necessary.

If both of you understand the complaint, let the other person know how you feel about the criticism. Calmly tell the person that you feel hurt, sad, or however you feel. Then tell the person that you are willing to work together to find a solution to the problem.

Following are some suggested steps for resolving conflicts. These steps may at first appear too structured. You will have to review and practice these steps to follow them. After a while, it will become second nature for you to solve conflicts this way. Go over the steps now as you imagine a situation between you and another person.

STEPS FOR RESOLVING CONFLICT

1) Invite the other person to talk with you about a problem you are having in your relationship.

2) State your complaint specifically. Make your feelings known and be sure to stick to the one issue. Assert yourself, using I statements.

3) To be sure your friend or partner understands your complaint, ask for feedback. If necessary, make clear what you meant.

4) Request a reasonable behavior change that serves both of your needs and solves the problem. Remember, the other person counts as much as you do.

5) Give your partner time to think about your request—at least a minute.

6) Allow the other person to express his or her feelings regarding your complaint or request. An open-ended question will help your partner tell you how he or she feels.

7) This time, you give your partner feedback about his or her feelings, but keep in mind step 8.

8) Take as much time as you need to think or calm down before replying. Avoid speaking out in anger. Remember that a workable agreement is not usually reached in a rush.

9) Allow your partner to (a) accept your request with no conditions or (b) accept your request under certain conditions. For example, you may request that you and your partner only double date for a while. This request may be accepted under the condition that you can see each other alone for half an hour after school on Wednesdays.

10) If the two of you cannot reach an agreement, it may be best to discuss the problem again later at a specific time.

Nonverbal communication
Body language

What about nonverbal communication?

The old saying "Actions speak louder than words" is especially true in relationships. Your **nonverbal communication** gives a message to others. Nonverbal communication is body language—the look on your face, your posture, or the way you move your hands, cross your legs, or dress. In a sexual relationship, the couple's touching and sounds—heavy breathing, sighs—can play a major role in communicating needs, wants, and desires.

When you communicate with your friends, teachers, a partner, or parents, make certain that your verbal and nonverbal messages match. Otherwise, you send mixed messages. Verbal and nonverbal communication should reinforce each other.

What do you think?

The degree of self-disclosure you use with different people often determines how close your relationship to them is. Think about three people you know: (1) one to whom you are close, (2) one to whom you would like to be close, and (3) one you have known for a while but to whom you aren't close. You may include friends, relatives, other adults, or anyone you wish. Rate on a scale of one (a little) to ten (a lot) how much you share with each person. Ask yourself why you self-disclose more with some people and less with others.

How will you decide?

Explore ways you behave aggressively, nonassertively, or assertively by monitoring your activities and feelings for three weeks. Keep a record of how you deal with certain people in different situations. Record times you behave aggressively, nonassertively, and assertively.

Choose areas to improve in behaving assertively rather than nonassertively or aggressively. Be sure to choose and note your strengths. Pat yourself on the back for situations in which you already behave assertively.

Evaluate why you are behaving nonassertively or aggressively in certain situations. Ask yourself why you are able to behave assertively in certain situations. This evaluation will help you reinforce your strong points and strengthen your weak points. Ask for feedback from someone you trust.

When do you find yourself behaving in nonassertive ways? Do you hesitate to offer comments in a classroom discussion? Do you think of the perfect response to someone who has hurt your feelings only after the incident is over? Or, do you tend to behave aggressively, bullying those around you so that you can get your way or to avoid being bullied first?

Decide on two strategies for becoming more assertive or less aggressive, and determine exactly when you will begin your efforts. Also decide on a specific date to reevaluate and, if necessary, improve your strategies. Note this reevaluation date on your calendar and stick to it.

Chapter 5 Review

Ideas to Remember

- Communicating well is one of the most necessary yet difficult responsibilities in any relationship.

- Communication about sexuality is especially difficult. Many people have little experience discussing this topic openly, or they may not know how to put the topic into words. Some parents may be frightened of exposing their children to sexual issues.

- Self-disclosure and trust are keys to developing friendships, especially intimate friendships.

- An intimate relationship is not necessarily sexual, and a sexual relationship is not necessarily intimate.

- Communication with parents can be beneficial, especially as you become independent of them.

- If you want to make your parents more "askable," part of the responsibility is yours.

- Assertiveness means you stand up for your rights without violating another person's rights.

- Through assertive behavior and a plan, you can learn to say no in a way that is firm but not hurtful.

- Basic steps can be followed to resolve conflict so that a relationship is improved.

- To avoid sending mixed messages, your verbal and nonverbal communication should match.

Questions to Ask Yourself

1) Why is real communication essential to responsible sexual relationships?
2) What are some things that you can do to improve the communication between you and your parents?
3) What are the four steps involved in saying no?
4) What should you keep in mind if you are going to criticize someone?
5) How can nonverbal communication help or hinder the communication process?

Action for Health

TALKING BLINDLY

This activity will show how much we depend on nonverbal behavior. Work with a close friend or relative and together choose a topic that is somewhat emotional between you. Find a timer that will let you know when ten minutes pass. Then sit back to back and begin a discussion of the topic you have chosen. Pay attention to your movements, posture, and expressions as the discussion becomes more intense. At the end of the ten minutes, turn around to face one other. Talk about any frustration you felt during the discussion because you couldn't see one another. How important is nonverbal communication?

STEPPING THROUGH CONFLICT

The ability to settle conflict is necessary for maintaining a healthy relationship. To practice this skill, work through this activity.

First, choose a friend or parent and a topic over which you have a minor disagreement. (You may wish to continue or restart the conversation you began in the Action for Health activity above.) It might help to review the steps for resolving conflict on page 86.

Now, one of you should raise the topic of conversation and assume the role of the speaker. The other should assume the role of the listener. Deliberately work your way through each of the conflict resolution steps, making note of problems you may have with them. When you have finished, talk about the difficulties you faced during the process. Talk about what can be done to overcome the difficulties.

Reverse roles and try it again.

Chapter 5 Review Sheet

Name _____ Date _____ Period _____

Read the imaginary conversation below. Then answer the questions that follow.

Hector: I suppose that you wouldn't go out with me even if I were the last guy on earth.

Lydia: I barely even know you!

Hector: So I'm giving you a chance to get to know me.

Lydia: I like to know the people I date a little better before going out with them.

Hector: I've heard that you like to play hard to get. I guess it's true.

Lydia: If you mean I'm picky about the people I date, you're right—I am.

Hector: So you think I'm hopeless.

Lydia: I'm glad you're interested, but instead of a date, how about eating lunch together? We'll see about a date later.

1) Read Hector's first line. List two communication don'ts in this sentence.

_____

_____

_____

2) Which of the characters uses self-disclosure? Which character plays a role? Support your answer with information from the passage above.

_____

_____

_____

3) Would you say that Hector and Lydia have mutuality? Support your answer with information from their conversation.

_____

_____

_____

Consider the conversation you have just read and remember what you learned from the chapter as you answer the following questions.

4) Explain why Lydia used assertiveness in her approach to Hector.

_____

_____

_____

5) Hector criticized Lydia for playing hard to get. How did she handle the criticism?

_____

_____

_____

6) Is Hector assertive or aggressive? Explain your answer.

_____

_____

_____

7) Imagine you are Hector. Write two sentences you would use to ask Lydia for a date. Refer to the text for good communication skills.

_____

_____

_____

8) Imagine you are Lydia. Use self-disclosure to explain how you feel about Hector's request for a date.

_____

_____

_____

Chapter 5 Critical Thinking

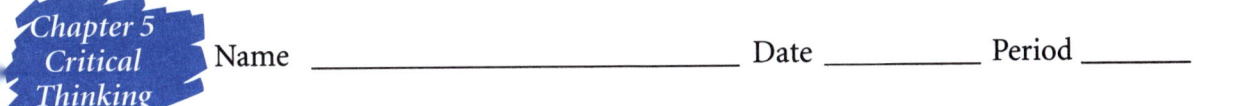

Name _____ Date _____ Period _____

The passage below is from "Bernice Bobs Her Hair," a short story by F. Scott Fitzgerald. The main character, Warren, has reluctantly agreed to dance with Bernice, who is visiting from Eau Claire, Wisconsin. Read the passage carefully. Then answer the questions that follow.

Warren danced the next full dance with Bernice, and finally, thankful for the intermission, he led her to a table on the veranda. There was a moment's silence while she did unimpressive things with her fan.

"It's hotter here than in Eau Claire," she said.

Warren stifled a sigh and nodded. It might be for all he knew or cared. He wondered idly whether she was a poor conversationalist because she got no attention or got no attention because she was a poor conversationalist.

"You going to be here much longer?" he asked, and then turned rather red. She might suspect his reasons for asking.

"Another week," she answered, and stared at him as if to lunge at his next remark when it left his lips.

Warren fidgeted. Then with a sudden charitable impulse he decided to try part of his line on her. He turned and looked at her eyes.

"You've got an awfully kissable mouth," he began quietly.

This was a remark that he sometimes made to girls at college proms when they were talking in just such half-dark as this. Bernice distinctly jumped. She turned an ungraceful red and became clumsy with her fan. No one had ever made such a remark to her before.

"Fresh!"—the word had slipped out before she realized it, and she bit her lip. Too late she decided to be amused, and offered him a flustered smile. Warren was annoyed.

Though not accustomed to have that remark taken seriously, still it usually provoked a laugh or a paragraph of sentimental banter. And he hated to be called fresh, except in a joking way. His charitable impulse died and he switched the topic.

Communication and Relationships Chapter 5

"Jim Strain and Ethel Demorest sitting out as usual," he commented.

This was more in Bernice's line, but a faint regret mingled with her relief as the subject changed. Men did not talk to her about kissable mouths, but she knew that they talked in some such way to other girls.

"Oh, yes," she said, and laughed. "I hear they've been mooning round for years without a red penny. Isn't it silly?"

Warren's disgust increased. Jim Strain was a close friend of his brother's, and anyway, he considered it bad form to sneer at people for not having money. But Bernice had had no intention of sneering. She was merely nervous.

Consider the passage you have just read and remember what you learned from the chapter as you answer these questions.

1) In your opinion, did the characters in the passage use self-disclosure? Use a sentence from the excerpt to support your answer.

_____

_____

2) In Chapter 5, you read some don'ts for communicating well. List two of these don'ts that Bernice and Warren used.

_____

_____

3) Write an I statement that one of the characters could use to express more accurately how he or she is feeling.

_____

_____

4) List two factors that make communication between adolescents difficult at times. Give an example from your own experience to support your answer.

_____

_____

Chapter 6

Controversial Issues

- ▶ What is family planning?
- ▶ What are contraceptive methods?
- ▶ What are some reasons for deciding against?
- ▶ What are some nonprescription and prescription contraceptive methods?
- ▶ What are other methods of conception control?

Janet and Prentice were taking another look at the "To Do" list for their upcoming wedding. Now that the ceremony was planned, they returned to their discussion about more personal matters—like when to start a family. Janet thought it best to wait a few years until her job brought in enough money.

"It might take more than five years before we're financially secure," Prentice said. "I'm not so sure I want to wait more than a couple of years."

"Well, let's just see how things go with our jobs. We can discuss it every now and then," Janet reassured him. "In the meantime, we will want to decide how to prevent having children before we're ready."

"We could just keep on doing what we're doing now—no sex," joked Prentice.

Janet laughed and replied, "Yeah, right. I think we better explore all the options, and then make a real decision about what's best for us."

What is family planning?

Family planning means that a couple plans when to start a family. They consider issues such as the strength of their marital relationship. They also consider the money needed for pregnancy, delivery, and raising a child. They think about the lifestyle they want, and how their careers could be affected.

In the past, this type of family planning was unworkable. Reliable methods of contraception—the prevention of conception and pregnancy—were unavailable. Today, family planning is quite easy because so many options are available.

Why is family planning important to adolescents?

Adolescent parenthood is increasingly common today. But it presents many problems to adolescents, parents, and society. For a young adolescent mother, pregnancy, birth, and child rearing bring personal, social, and biological difficulties. Babies born to young teenagers are twice as likely to have low birth weight as babies born to older females. These babies are also two to three times as likely to die within the first twenty-eight days of life. Also, illness and hospitalization rates are higher for babies born to adolescent mothers.

How does a pregnancy fit into your life goals and plans right now?

While not all teenagers who fail to delay pregnancy and childbearing have negative experiences, research shows that:

- Adolescent mothers frequently have to depend on others for financial help.
- Adolescent mothers often lack adequate parenting skills that time, education, maturity, and experience can help provide. Often grandparents end up caring for infants.
- Costs to society to help adolescent mothers are great. In 1993, 159,000 teenage mothers ages 18 to 19 and 32,000 ages 15 to 17 received Aid to Families With Dependent Children. This money comes from taxpayers and consumers, who end up paying high medical premiums.
- Adolescent mothers are at much greater risk of having a second child during their teenage years than are teens who have not yet had a child.

> **Abortion**
> The ending of a pregnancy

- Teenage fathers are generally not prepared for the responsibilities of parenthood. They are at a greater risk for academic, drug, and behavioral problems than are other adolescent males.
- Nearly one-third of all adolescent pregnancies end through **abortion**—the ending of a pregnancy.
- Adolescents who have sexual intercourse without contraception are at risk for pregnancy and sexually transmitted disease. Their risk is greater than for teens who regularly use an effective contraceptive method.

What does a contraceptive method do?

A contraceptive method prevents the joining of an egg and a sperm and, therefore, prevents conception and pregnancy. Contraceptive methods are divided into the following categories:

- abstinence—not having sexual intercourse—the ideal method for adolescents
- nonprescription methods, which do not require a visit to a doctor or a prescription
- prescription methods, which do require a visit to a doctor and a prescription
- sterilization, which requires surgery

What are some reasons for deciding to abstain?

The choice to abstain, or not to have sexual intercourse, shows responsible decision making. As an adolescent in today's world, you may have to make a decision often about sexual intercourse. As with Prentice and Janet, the decision to wait is easier for those who remember the potential consequences of having intercourse. Even if a person has said yes one or more times, the next decision can be to say no. Many facts support the decision to wait. Here are some of them:

- As an adolescent, you are still maturing in your sexual development, sexual identity, and sexual relationships.
- Sexuality means much more than sexual intercourse.

- Waiting allows the time to learn to communicate effectively, honestly, and deeply.
- Time allows you to see your partner as a friend and not just as a sexual object.
- You can learn to set and enforce limits that may also be helpful in the nonsexual areas of your life.
- Couples can learn to solve problems together as a team.
- As you are maturing, you can learn how to have pleasurable, even sensual experiences that do not lead to sexual intercourse.
- You can develop true intimacy through meaningful self-disclosure rather than a false intimacy based only on physical disclosure.
- You will avoid worry about getting sexually transmitted diseases that could lead to cancer or AIDS, which leads to death.
- Abstinence will relieve you from concern about finding and using a contraceptive method.
- Abstinence will relieve you from concern about getting pregnant or making someone pregnant.
- You can learn to express your love in different ways that will be a great addition to a marriage.
- You won't have the concern that "everyone knows you're doing it."
- You won't suspect that your relationship is continuing simply because you are having sexual intercourse.
- The possibility of making a painful decision is removed, such as giving up a child for adoption or whether to end a pregnancy.
- You will have no fears about your parents "finding out."

Keep in mind that deciding to wait does not mean you will never have physical contact or intimacy. It does mean that you have made a responsible decision to allow you, your partner, and your relationship to unfold gradually.

Withdrawal
An unreliable contraceptive method involving the removal of the penis from the vagina before ejaculation

Douching
Flushing the vagina with liquid; sometimes used as a contraceptive method

Male condom
A thin sheath, usually made of latex rubber, that is worn over the penis to stop sperm from entering the vagina

What are some nonprescription methods of conception control?

Some contraceptive methods are effective, and others are not effective at all. The following methods do not require a prescription, meaning a they can be used without a doctor's written order. If any but the first method is chosen, all directions must be read and followed completely and carefully.

Withdrawal is the removal of the penis from the vagina before ejaculation. The method is unreliable because it is common for a drop of fluid containing thousands of sperm to escape from the penis before ejaculation. Therefore, even if the male removes his penis before ejaculation, the chance of pregnancy is still good. Sperm anywhere near the vaginal opening may still make their way up into the vagina and fallopian tubes. Remember, a single ejaculation of semen has between 40 and 120 million sperm. And it only takes one sperm and one egg to create a pregnancy. Finally, withdrawal can be frustrating for both partners. In short, withdrawal is not an effective way to prevent pregnancy or an STD infection.

Douching, is flushing the vagina with liquid. It too is ineffective at preventing pregnancy. Douching with a spermicide, a chemical that kills sperm, is also ineffective. By the time the female douches, it already may be too late to prevent fertilization. Douching is better than doing nothing, but not much.

A **male condom** is a thin sheath, usually made of latex rubber, that is worn over the penis to stop sperm from entering the vagina. It looks much like the finger on a rubber glove. Many condoms have a small nipple-like space at the tip to hold the ejaculated semen. A condom is unrolled onto the erect penis before insertion into the vagina. Lubricants such as petroleum jelly tend to weaken condoms, causing them to break. Prelubricated condoms are much safer. The best condoms to use have spermicide as a lubricant, which gives added

unrolled condom

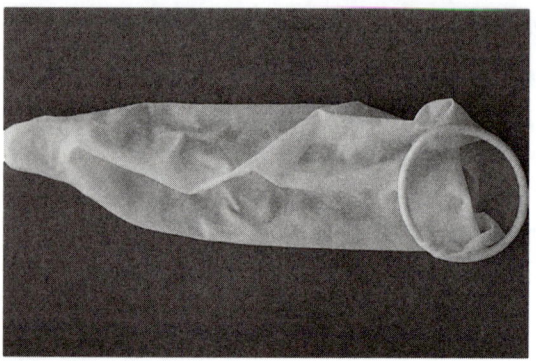

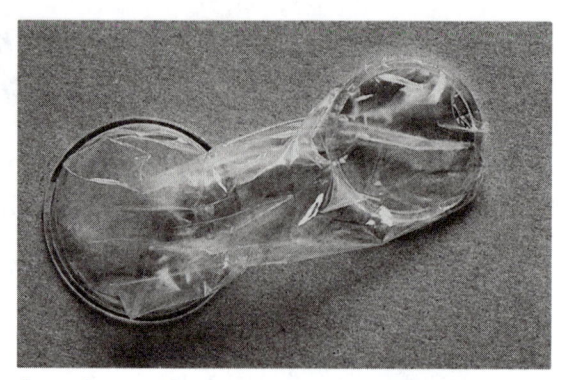

female condom

Female condom
A prelubricated latex contraceptive device that completely lines the vagina to prevent sperm and sexually transmitted diseases from entering

Vaginal spermicide
A contraceptive such as a foam, cream, jelly, suppository, or sponge that is inserted into the vagina to destroy sperm

protection against pregnancy if the condom breaks. A condom should be removed carefully to prevent semen from spilling. Each time a couple has intercourse, a new condom should be used. In addition to preventing pregnancy, condoms can prevent the spread of sexually transmitted diseases, including HIV. However, only latex condoms are effective in protecting against STDs. Condoms made from animal tissue, often called skin condoms, do not protect the wearer against STD viruses. Condoms should be stored in a cool, dry place. They are most effective when used properly with another contraceptive method.

The **female condom** is a prelubricated latex contraceptive device that lines the vagina completely to prevent sperm from entering. Like the male condom, this method protects not only against pregnancy but also against sexually transmitted diseases. The female condom also can be used only once and should be removed carefully before standing to prevent spilling.

Vaginal spermicides come in several different forms, such as foam, creams or jellies, suppositories, and the contraceptive sponge. When used with a condom, spermacides can be very effective. Foam is the most widely used vaginal spermicide because of its safety, effectiveness, and availability. Foam comes in pressurized cans, often with prefilled, disposable applicators. Foam should be inserted no more than thirty minutes before sexual intercourse.

100 *Chapter 6 Controversial Issues*

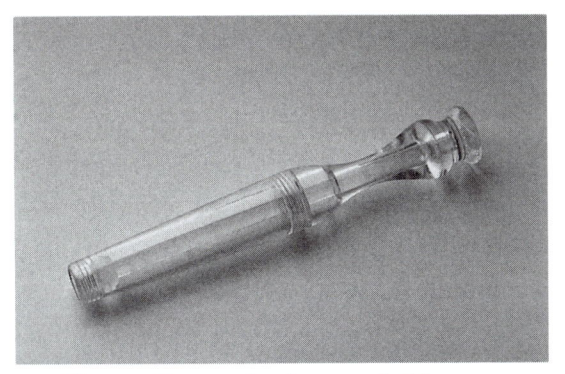

applicator for vaginal spermicides

contraceptive sponge

> **Toxic shock syndrome (TSS)**
> *An infection caused by bacteria present in the vagina during menstruation*

Spermicidal creams and jellies function in a similar manner to foam. Like the foam, creams and jellies should be inserted no more than thirty minutes before intercourse. Foam, creams, or jellies must be left in place without douching for several hours after intercourse.

Spermicidal suppositories are tablets that dissolve after being inserted deep into the vagina. The couple should wait at least ten minutes but not more than an hour to have intercourse.

The contraceptive sponge is a soft, white, round sponge that contains spermicide. The sponge must be moistened with a little clean water and then squeezed until it foams to activate the spermicide. After being inserted deep into the vagina, the sponge traps, absorbs, and kills sperm. It should be left in place at least six hours after intercourse. A word of caution about the contraceptive sponge. It has been linked to **toxic shock syndrome (TSS)**. TSS is a serious infection caused by bacteria present in the vagina during menstruation. Although TSS rarely occurs, precautions should be taken. For example, a woman should not use the contraceptive sponge during menstruation. She should also be certain to wash her hands carefully with soap and water before insertion and removal of the sponge. The sponge should be left in at least six hours but not more than twenty-four hours.

Controversial Issues Chapter 6

Natural family planning
A series of conception control methods based on determining when a woman ovulates

Oral contraceptives
Pills taken daily by mouth to prevent conception

Diaphragm
A contraceptive device that fits over the cervix, working as a barrier to sperm

Intrauterine device (IUD)
A small, T-shaped object inserted into the uterus to disrupt the normal chemistry of the uterine lining

What are other nonprescription methods of conception control?

Some other contraceptive methods also don't require a prescription. **Natural family planning** is a term that refers to a series of conception control methods. You may recognize such names as the rhythm method, the safe period, or the calendar method. All of these methods have a single goal—to determine when a woman ovulates. Knowing when ovulation occurs helps a couple pinpoint the days when a woman is likely to become pregnant. To use natural family planning successfully, a couple must then abstain from sexual intercourse on those days.

The different names of these techniques tell how they attempt to find out when a woman is ovulating. For example, the calendar method requires that a couple chart the woman's menstrual cycle for one year to determine her longest and shortest cycles. A formula is then used to calculate the "safe" days, or days when she is not likely to become pregnant. Natural family planning requires careful record keeping and charting as well as unwavering discipline—something difficult even for older, more experienced couples.

Clearly, there is no simple way to find out a safe time to have sexual intercourse. In fact, for many adolescent girls whose cycles are irregular, there is no safe time of the month to have intercourse without risking pregnancy.

What contraceptive methods require a doctor's prescription?

If Janet and Prentice decide to visit a physician, three contraceptive methods are most likely to be discussed—**oral contraceptives** (the "pill"), the **diaphragm**, and the **intrauterine device (IUD)**.

Oral contraceptives Also called birth control pills, oral contraceptives are taken daily by mouth to prevent conception. They contain hormones similar to the natural female hormones estrogen and progesterone. The amount and release of these hormones varies according to the kind of pill.

To prescribe a particular type of pill, a physician would need to talk with Janet and give her a physical examination. The doctor would ask her about several health factors, including any family history of diabetes and heart problems. The doctor also would ask about her current smoking status. Based on these factors, the doctor would decide whether the pill is suitable for Janet. If so, the doctor would prescribe the appropriate pill.

If Janet and Prentice choose the pill, she would have to take it around the same time every day. They would also need to be aware that if Janet is given another drug for an illness or infection, it could lower the effectiveness of her birth control pill. If Janet has any side effects, her physician possibly could prescribe a more suitable pill for her.

Diaphragm The diaphragm is a round, soft latex dome with a flexible latex-covered spring around the rim. Diaphragms, which fit over the cervix, come in a range of sizes and styles. Janet's physician would fit her with a diaphragm appropriate to the size of her cervix. A small amount of spermicidal cream or jelly must be placed into the cup of the diaphragm and around the rim before it is inserted into the vagina. The diaphragm works as a barrier to sperm, and the spermicide destroys the sperm. Most women find the diaphragm comfortable, and a couple usually cannot detect its presence during intercourse. Because the diaphragm also has been associated with toxic shock syndrome in some women, it is important to discuss it with a doctor.

Intrauterine device Today's intrauterine device (IUD) is a small, flexible, T-shaped object that is inserted into the uterus. It is believed that the presence of

open package of birth control pills

diaphragm

> **Sterilization**
> A surgical procedure that makes a person incapable of conception
>
> **Vasectomy**
> A sterilization procedure for men that blocks the passage of sperm

the IUD irritates and disrupts the normal chemistry of the uterine lining. Once inserted, the IUD may remain in the uterus for up to several years. When a woman wants to become pregnant, she has the device removed.

If Janet and Prentice seek this method of conception control, Janet must see her physician. Her doctor may prescribe one of two types of IUDs. One type contains the female hormone progesterone. The other type of IUD has copper bands on the horizontal arms of the IUD and copper wound around its stem.

Although the IUD has had a controversial history, these two devices now provide a safe alternative contraceptive choice. Physicians are very selective in choosing which of their patients should use the IUD. A general guideline is that it is safest to prescribe the IUD for older women who usually have already started or completed a family.

What is sterilization?

Sterilization refers to a surgical procedure that makes a person incapable of conception. It is almost always permanent. While sterilization is a popular contraceptive method for some older people, it may not be a wise choice for younger people. Younger women and men are less likely to have started or completed a family. Because Prentice and Janet eventually want a family, they won't choose this method.

A **vasectomy**, shown in Figure 6.1, is a sterilization procedure for men that creates a block against the passage of sperm. It is a simple procedure that can be performed quickly and safely in the physician's office. The doctor makes a small incision in each side of the scrotum and then cuts, ties, and often seals the vas deferens. This minor surgery is usually completed in about twenty minutes, and the man can return home in a half hour. After a vasectomy, a man still produce sperm, but they are blocked and the body absorbs them.

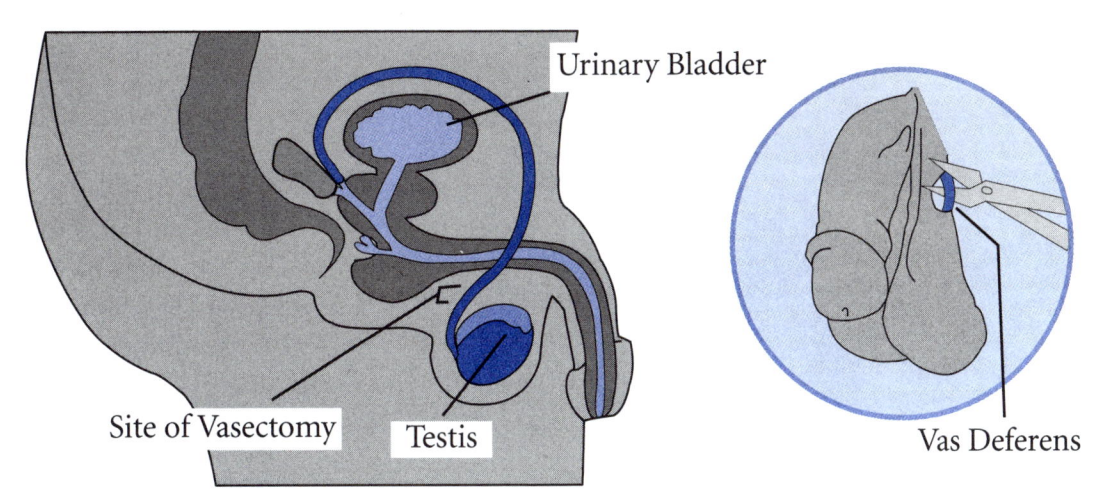

Figure 6.1 Vasectomy

Tubal sterilization
A sterilization procedure for women that blocks the fallopian tubes

Tubal sterilization involves blocking a woman's fallopian tubes. Doctors may use several different methods for causing this blockage. In one method, the fallopian tubes are cut and tied as shown in Figure 6.2. In another method, clips or rings are used. This type of blockage provides the best opportunity for the process to be reversed. After any type of tubal sterilization, a woman still releases eggs, but they are blocked, and the body absorbs them.

What should be considered when choosing a contraceptive method?

The couple involved must decide whether to use contraception and what method they will use. If they prefer a prescription method, they must seek a physician's advice. Janet and Prentice are moving in the right direction by exploring all the options and then discussing them. Individuals who are about to engage in intercourse or choose a contraceptive technique need to consider questions around these three main issues:

Personal Lifestyle Issues

- Should I wait until I am married to have intercourse?
- How often will I be having sexual intercourse?
- How important are children to my partner?

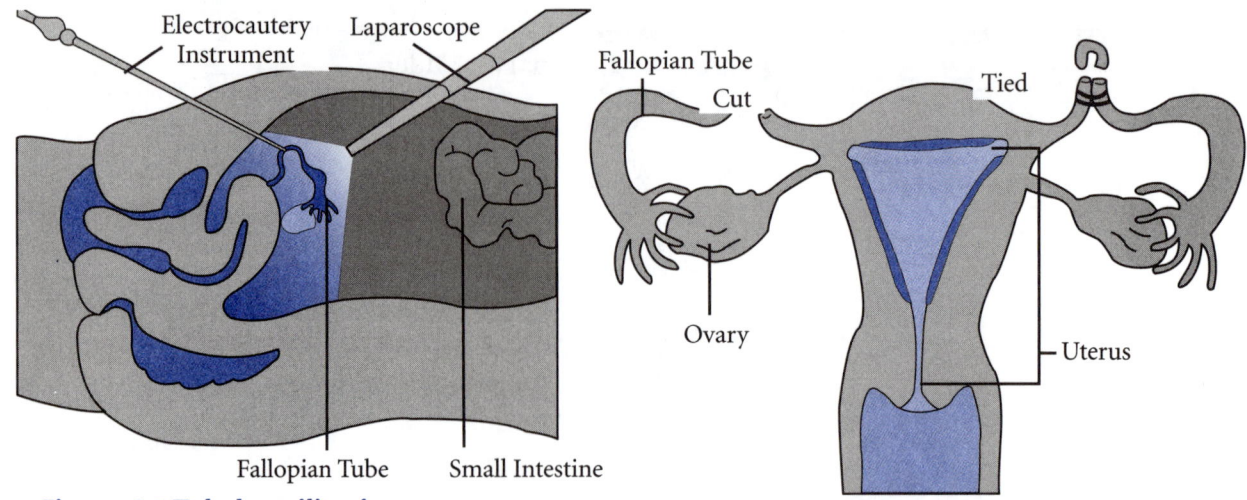

Figure 6.2 Tubal sterilization

- Do I want to have children? If so, when do I want to have my first, or next, child?
- Can I financially support one child or more?
- Would I remarry and want more children if a first marriage ended?
- What obligation do I have to limit my family size?

Personal Health Issues

- What is my present health status?
- Do I, or does my partner, smoke?
- For the woman: Is there a history of diabetes or heart disease in my family?
- Did my doctor list any side effects of a method that apply to me or my partner?

Contraceptive Technique Issues

- How effective is each of the techniques?
- How safe is each of the techniques?
- How expensive is each of the techniques?
- When, how, and how often is each of the techniques applied?
- Which technique am I most likely to use faithfully?

Remember that abstinence is the most effective, safest, and least expensive way to prevent pregnancy and to control the spread of sexually transmitted diseases.

Why is abortion a hot topic?

As you are probably aware, abortion is a highly controversial issue. When an abortion occurs unexpectedly for known or unknown reasons, it is called a miscarriage. A couple has little or no control over a miscarriage. In contrast, a qualified physician purposely brings about a surgical or medical abortion in a medical setting. Some physicians will perform abortions while others won't. Some states have strict laws about abortion, and others are more liberal. People who favor a woman's right to choose abortion believe that a woman has a right to control her body. People who oppose abortion, or prolife supporters, believe that the fetus has the right to life.

Those who act responsibly are fully aware of the potential consequences of their actions and avoid high-risk behaviors. Preventing an unplanned pregnancy eliminates even considering abortion.

What are your opinions on abortion as a means of terminating an unwanted pregnancy?

What do you think?

In a group or as a class, briefly review the methods of conception control covered in this chapter. List or chart and discuss the role each gender plays in acquiring and using each method. In your list or chart, include:

- steps involved in acquiring each method
- who could or does accomplish these steps
- the risks involved in using each method
- who takes each of these risks
- the costs of each method
- who might pay for each method and how

Afterward, discuss the benefits and drawbacks of taking responsibility for a conception control method. Finally, brainstorm ways that both partners can take more equal responsibility for each method.

How will you decide?

Explore the future you will make for yourself by plotting life events on a timeline. You don't need an exact life plan in mind—this is just an exploration. In making your timeline, allow yourself to daydream. Before you begin, write answers to these questions.

1) How old will you be in five years?

2) How would you like to be supporting yourself at that time?

3) What will you do in your leisure time, and what will your friends be like?

4) What kind of person do you want to be in five years?

5) What changes and events will have happened by then—college? pursuit of a career? travel? several dating partners?

Design a timeline with increases of your choosing. Go out at least ten years.

Choose important events or processes to plot on your timeline. For example, plot marriage, children, or any other important events you imagine to be part of your future.

Evaluate how each of the plans, events, and ideas would change if you became a parent one year from now. Note the child's age when each event on your timeline would take place.

Decide how you will delay or avoid pregnancy in your future. Indicate on your timeline the years you will use your chosen conception control. If marriage is in your immediate future, plot the age at which you feel you'll be financially and emotionally able to care for a family.

Chapter 6 Review

Ideas to Remember

■ Family planning is a couple's effort to decide when and whether to start a family.

■ While childbearing may be a positive experience for a few adolescents, most teenage parents experience serious negative consequences.

■ There are several reasons for delaying sexual intercourse, many of which involve your degree of sexual maturity and your sexual identity.

■ Withdrawal and douching are not considered effective methods for preventing pregnancy.

■ There are prescription, nonprescription, and surgical methods of conception control.

■ Sterilization is a surgical procedure that permanently prevents pregnancy.

■ Abortion is a highly controversial procedure used to terminate an unwanted pregnancy.

Questions to Ask Yourself

1) How would you explain "family planning" to a classmate?

2) How does family planning fit into your current lifestyle? How might it fit into your future lifestyle?

3) What are some nonprescription contraceptive techniques available to adolescents?

4) What are some prescription methods a physician is likely to discuss with an adolescent seeking contraception?

5) How does sexually responsible behavior fit into the scheme of family planning and abortion?

Action for Health

HELPING MARIA DECIDE

The decision to delay sexual intercourse can be difficult. Read the paragraphs below, answer the questions that follow, and then continue the activity.

Maria, a senior, lives with her mother, her brothers, and her sisters in a small apartment in the city. She has been dating Umberto for six months and feels she loves him very much. Umberto, who graduated from high school last year, has a steady job as a restaurant cook.

Maria decides to see her counselor. She explains that she and Umberto have had sex only one time. She knows that he will persist about having it again soon. Maria's religion, which is important to her, tells her she should not have sex before marriage. And, her mother would be upset.

Maria believes that if she did become pregnant, Umberto would marry her. That would make her happy. She could move out of the house and start a life of her own. She would get a job and maybe even be able to give her mother a little money each month. Maria feels she should probably not have sex again. On the other hand, she loves Umberto and doesn't want to lose him.

1) What are the major issues presented in this situation?

2) If you were Maria's counselor, what are some questions you would ask her?

With another student or your teacher, role-play Maria's asking her counselor for advice. Discuss all appropriate details. Have Maria and the counselor agree on a positive step she can take. You could continue the activity by role-playing the next meeting between Maria and Umberto or one between Umberto and the counselor.

Chapter 6 Controversial Issues

Chapter 6 Review Sheet

Name _____ Date _____ Period _____

Part A Match the description in the left column with the sexual body organ in the right column. Write the letter of the correct answer on the blank.

_____ 1) contraception
_____ 2) family planning
_____ 3) contraceptive method
_____ 4) abstinence
_____ 5) sterilization
_____ 6) spermicide
_____ 7) male condom
_____ 8) foam
_____ 9) toxic shock syndrome
_____ 10) natural family planning
_____ 11) oral contraceptives
_____ 12) diaphragm
_____ 13) vasectomy
_____ 14) medical abortion
_____ 15) female condom

a) not having sexual intercourse
b) chemical that destroys sperm
c) latex dome that covers the cervix
d) pregnancy is terminated by a physician
e) technique to prevent the joining of an egg and sperm
f) vaginal spermicide
g) prevention of conception and pregnancy
h) hormone pills that prevent pregnancy
i) choosing when to start a family
j) surgery to prevent conception
k) thin sheath to stop sperm from entering vagina
l) male sterilization procedure
m) infection caused by bacteria present in the vagina during menstruation
n) prelubricated device that lines the vagina
o) series of methods that prevent conception by determining when a woman ovulates.

Part B Explain how the following contraceptive methods work.

1) Vaginal spermicides

2) Contraceptive sponge

3) Natural family planning

4) Withdrawal

5) Male condom

6) Diaphragm

7) IUD

8) Female condom

Chapter 6 Critical Thinking

Name _____ Date _____ Period _____

The letter below was written to Abigail Van Buren, a newspaper advice columnist. Read it carefully. Then answer the questions that follow.

Dear Abby,

I am writing this as a follow-up to the letter I wrote you in June about my 13-year-old daughter.

(I told you she was getting dangerously close to the boyfriend she had been seeing morning, noon and night.)

I took your advice and spoke to her about sex. I purchased at the local drugstore several forms of birth control (condoms, contraceptive sponge and vaginal suppositories). I also bought something called "teen pack," which contained several trial-size items introducing young females to such things as tampons, mini-pads, maxi-pads, shaving lotion and razors for shaving legs. When I got home, I invited my daughter to join me at the kitchen table. My husband was working late that night, so it was a perfect opportunity for girl talk.

I lined up the contraceptive devices on the table. My daughter was a bit curious. I gave her the teen pack, telling her that all the items in that package were things that we had discussed. Then I pointed to the other items on the table and said they were items that we needed to talk about.

I carefully explained to her that now that she had become a young woman having a monthly period, she could become pregnant if she had sex. I then told her that I was in no way condoning sex in someone as young as she, but that I wanted her to be informed. I then took each product and explained how it was used, and showed her the directions and how to check the expiration date on the product. After that, I opened each package and let her touch the device, examine it and ask questions.

I made sure she was aware that even if you are on the birth control pill, that still did not stop sexually transmitted diseases such as AIDS. I told her that safe sex with condoms could not completely guarantee the prevention of pregnancy or disease, but that the only way to ensure not getting pregnant or a sexually transmitted disease was not to have sex.

I then took the remaining products and placed them in a box in her bathroom closet. I told her I would not check the box, but if she ever felt a need to experiment, that they would be there. I stressed to her to always feel that she could come to me before making any decisions that could change her life forever.

And I made some important points to her that truly hit home: I pointed out that she was too young to obtain a driver's license, too young to drink, too young to get a job, and too young to be responsible for the life of another human being. I reminded her that if she were to have a child, her education and social life would cease until such time that she could afford a baby sitter to resume her studies and social activities.

Consider the excerpt you have just read and remember what you have learned from the chapter as you answer the questions below.

1) Re-read the last paragraph of the passage on the previous page. Use your text to list two more reasons for delaying pregnancy that adolescents should consider.

   _____
   _____

2) List four methods of contraception the mother did not mention.

   _____
   _____

3) Write a sentence explaining why you think adolescents should be informed about family planning.

   _____
   _____

4) In your opinion, was the mother's technique with her daughter effective? Explain your answer.

   _____
   _____

114 *Chapter 6 Controversial Issues*

Chapter 7: Sexual Exploitation and Harassment

- ▶ What are sexual exploitation and harassment?
- ▶ What can be done about sexual harassment?
- ▶ How does rape differ from sexual harassment?
- ▶ What should a rape victim do?
- ▶ What is incest?

Russell was explaining to Dr. Copas, a psychiatrist, how one of his college professors was sexually harassing him. Frequently the professor tried to touch or hug him. She often suggested they "get together" and hinted that Russell's grades might depend on his decision. She implied that even if he told someone, no one would believe him because she is a "well-respected professor." Russell hadn't slept for weeks. He hoped Dr. Copas would give him something to help him relax. Whitney, a rape counselor, was listening intently to Jennifer, a 15-year-old girl. Jennifer's 17-year-old boyfriend had raped her. Jennifer tearfully told how her boyfriend had forcibly held her down and called her names. She had tried to resist, but he forced her to do sexual things.

Both Russell and Jennifer were victims of sexual assault. Although they both felt disgraced and degraded, they had found the courage to seek help.

> **Sexual exploitation**
> Taking advantage of another person using sexual activity as a weapon
>
> **Sexual harassment**
> Exploitation using power or status to force someone into sexual behavior

What does sexual exploitation mean?

Sexual exploitation, or abuse, occurs when one person takes advantage of another person. The weapon in this exploitation is sexual activity, and the victim's feelings or rights are not considered.

In loving relationships, the partners use sexual activity to show how much they care about each other. Sexuality can be a means of sharing your deepest, most profound emotions. However, sex also can be used to humiliate, or degrade, another person.

Sexuality and sexual activity may be good or bad, depending on how they are used or abused. For example, if someone you like kisses you, that might be nice. On the other hand, if someone you don't know or don't like much kisses you, that could be upsetting or even disgusting. Your feeling about the kiss depends on who is doing the kissing as well as how and why it is done.

Unfortunately, some people use sexual activity to take advantage of others. Such sexual exploitation is not usually done for sexual reasons. Instead, the reasons can range from a wish to show power to a desire to humiliate another person. Some common ways that people may be sexually exploited are through harassment, rape, or incest.

What is sexual harassment?

Sexual harassment is exploitation using power or status to force someone into sexual behavior. It occurs in many forms and in many different places. In a work setting, an employer, superior, or coworker may use the power of threats or promises to take sexual advantage of someone. For example, a supervisor might threaten to deny a salary increase or promotion unless the employee grants the supervisor a sexual favor. In an educational setting, a teacher may hold the power of grades or references over a student to gain sexual favors.

Sexual harassment may involve suggestive stares, sexual comments and invitations, or various kinds of touching. In all cases, the victim is treated like an object rather than a human being with equal rights.

In most cases of sexual harassment, power is used to try to force someone into sexual behavior. The victim often feels as if the situation is hopeless. Recall Russell's situation. If he refuses to do what his professor suggests, he may fail the course. He feels this may affect his major and his job potential after school. On the other hand, Russell knows that if he does give in, he may never feel good about himself again. You can see there is much more to sexual harassment than just sexual activity. Sexual harassment may have lasting, damaging consequences.

What can be done about sexual harassment?

Sexually harassing someone is against the law. If you or someone you know is being sexually harassed, some of the specific suggestions listed in the box may help handle it.

HANDLING SEXUAL HARASSMENT

Seek personal support. Find a friend, support group, or counselor you can talk with about the harassment. Your supporters can help you see things clearly and realize that the sexual harassment is not your fault.

Let the offender know how you feel. You can tell the offender face to face. Or, you could write a letter detailing what has been done to you and how you feel about it. In either case, you can state that you want the harassment to stop immediately. If this approach doesn't seem to help, you may need to take further action.

Report the incidents to the proper authorities. In a work setting, you may need to report the harassment to your immediate supervisor or the human resources department. Check the company's policy on harassment to be sure of the procedure. In a school setting, you can report harassment to an administrator or guidance counselor. Your complaint may be one of several other complaints against the same person. This may begin an investigation and encourage other victims to come forward. Keep careful records to support your case.

Understand your role. As you deal with this problem, realize that you are a victim of someone else's behavior. You did not invite this behavior and are not responsible for the offender's actions.

Sexual Harassment

Rape
Sexual intercourse without consent and with actual or threatened force

How does rape differ from sexual harassment?

Rape is generally understood to mean sexual intercourse without consent and with actual or threatened force. By comparison, sexual harassment does not have to involve sexual intercourse. Statutory rape is sexual intercourse with a minor, which usually means a person under age 18. In cases of statutory rape, the attacker can be charged even if the minor consented.

In some rapes, sexual intercourse as we normally think of it does not occur. A person can be forced to have anal intercourse, which is commonly considered rape. Many states include such rapes under their sexual assault laws.

Who is most likely to be raped?

Rape can happen to anyone. However, statistics show that adolescent girls and women in their early twenties are at greatest risk of rape and sexual assault.

It is a myth that most rapes are committed by an unknown attacker in such places as dark alleys late at night. More often than not, a rape victim—especially an adolescent victim—

> **Date rape**
> *A form of rape in which the victim has been dating the attacker*

knows her attacker as an acquaintance, friend, dating partner, or lover.

Figures about male rape victims are scarce. Most male rape victims are average people. Unlike female rape victims, most male rape victims do not know their attackers.

What does date rape mean?

Date rape means that the victim has been dating the attacker. This is the most common form of rape among adolescent girls and women in their early twenties. In some cases, people may say that a young woman has encouraged an attacker. This is wrong. While some girls may act seductively, or as if they are tempting the male, they are not asking to be raped. Unfortunately, this is sometimes how victims are treated in courtrooms.

You may wonder if males are ever victims of females in date rape. Evidence from rape crisis centers shows that they are, although it seems to be rare. Sometimes a girl or woman threatens to tell the male's parents or other authorities something negative about him if he doesn't go along with her wishes. Young men may give in to the pressure to have intercourse just as some young women do. Sometimes it is out of fear that their relationship will end.

Some common factors in date rape situations are:

- Poor communication
- The mistaken idea that a woman says no but really means yes
- The female trusting her dating partner too much
- Use of alcohol
- The male's need to prove his manhood and power

None of these factors is an excuse for rape. In fact, there is never an excuse for rape. Regardless of the circumstances, sexual intercourse against someone's will or without his or her consent is rape. Keep in mind that rape is never the victim's fault.

What warnings signs can a young woman look for in a dating relationship?

Date rape cannot always be prevented, but young women can look for some of the following signals.

- Your dating partner makes physical advances against your wishes.
- He attempts to make you feel guilty about saying no.
- He shows lack of respect for you, for your feelings, or for women in general.
- He holds the belief that women are meant to serve men.
- He uses physical force to make you do anything that goes against your better judgment.
- He drinks heavily or abuses other drugs.
- He becomes verbally abusive or uses physical force while under the influence of alcohol or other drugs.

There are some ways to try to prevent date rape. On the next page, the Suggestions for Preventing Date Rape apply to girls and women, and they also help boys and men avoid becoming attackers.

What may increase the possibility of being raped?

Research shows that certain factors contribute to the likelihood of rape. More rapes occur against women who are known to be sexually active and live separately from their parents. A higher number of rapes occur among women who experiment with drugs or alcohol and do not have a "visible" boyfriend.

Research shows fewer rapes occur among girls who do not associate with a sexually active peer group or a group that uses drugs. Also, fewer rapes occur among girls who have good relationships with parents, teachers, and other adults.

There is no ideal plan for avoiding rape. If you find yourself in a rape situation, you may try talking your way out of it. Others have found that vomiting or acting insane has halted the assault. Some women struggle with their attackers and manage to escape. Many others have found that simply screaming is effective. If you are threatened with a weapon, resistance may not be the best action.

> **SUGGESTIONS FOR PREVENTING DATE RAPE**
>
> *Women*
> - Have good basic knowledge of sexuality.
> - Know what behaviors are morally right for you.
> - Before any date, know your limits and how you will enforce them.
> - Firmly but politely let your partner know your wishes.
> - Be aware of your nonverbal communication—how you dress and how you touch should match what you say.
> - Pay attention to where you are and what is going on.
> - Do not go off alone together. Let your friends or family know where you're going.
> - Do not use alcohol or other drugs—they will confuse your ability to think and to assert yourself.
>
> *Men*
> - Have a good basic knowledge of sexuality.
> - Know what behaviors are morally right for you and your partner.
> - Know your sexual desires and the true motives of your actions.
> - Communicate your interest and your limits to your partner.
> - Use self-discipline.
> - Know that a no answer to sex is not a rejection of you—just of the behavior.
> - Don't think that because a girl is attractive, she is willing to have sex.
> - Even if a dating partner has had sex with you before, it does not mean she is willing to do it again.
> - Do not use alcohol or other drugs, because they may cause you to become sexually irresponsible.
> - Know that sexual intimacy means much more than sexual intercourse.

In spite of all efforts and actions, there is always the danger that you may become a victim of rape. Remember that being raped isn't your fault. Rape is a product of power and violence, not sexuality.

What should a rape victim do?

If you or a friend becomes a rape victim, do the following:

- Contact the police.
- Do not shower, change clothes, or do anything else before going to the hospital for a physical examination. Doing otherwise can destroy important evidence.
- Contact a friend or parent for support.
- Work with a rape crisis or support group to help you deal with your many different feelings.

If a friend told you she had just been raped, what would you do? What would you suggest she do?

Incest

Incest
Sexual activity between family members, usually involving a child and an adult

What is incest?

Incest is sexual activity between family members, usually involving a child and an adult. Unlike rape, physical force is rarely used. Instead, the adult usually uses bribes or threats to persuade, trick, or deceive the child.

Most children who become victims of incest do so without realizing it. Often their loyalty to the adult or to the family keeps them from reporting the incident. In many cases, they may not even realize the behavior is unusual. Most incest victims develop a distorted view of sexuality. They lose their trust in family relationships.

Children who are incest victims should understand that it is not their fault. They need to see the importance of reporting the adult who has abused them. If their story isn't accepted at first, they must continue to try to make it known. Only then will they be protected from further abuse. In many cases of child abuse and incest, the adult was abused as a child. Reporting the incest may help the adult get counseling. It may even prevent the person from harming other children.

What would you tell a fourth-grade student who told you a family member was sexually abusing him?

What do sexual harassment, rape, and incest have in common?

As you have read this chapter, you may have noticed two common themes. First, violence and a desire for power, not sex, cause sexual harassment, rape, and incest. Second, at no time are the victims of sexual harassment, rape, or incest at fault.

What do you think?

Sexual exploitation is more about power and aggression than it is about sexuality. Reinforce what you have read in this chapter by considering the following points.

1) Think about your overall behavior and attitude toward the opposite sex. Do you treat them as people to be respected? Do you view them as a group somewhat better or worse than your own? Do you ever talk about them in negative terms? Explain.

2) Do the types of magazines, movies, and music you enjoy promote abuse and degrade women or men? Do they promote respect for the opposite sex? Share some examples with the class.

3) Are your motivations for engaging in sexual behavior driven by peer group pressure? by your own opinions? by a desire to improve the relationship? Write your honest responses.

4) Think about a sexual activity or a situation that would be disgusting to you. Imagine someone forcing you to take part in that activity. Would this incident be something you would forget? How do you think it would affect your future relationships?

How will you decide?

Explore, discuss, and list the precautions you and your classmates take to avoid becoming victims of violent crime. Also discuss any practices and behaviors that help to ensure the safety of your family and neighbors. Then list behaviors you see in other students, family members, or neighbors that put them at risk for becoming a victim.

Choose and list any precautions you are not using. Also list behaviors you may be practicing that put you at risk.

Evaluate how well you, your family, and your neighborhood are avoiding violent crime. Evaluate how you can improve this.

Decide to acquire one or more new safety practices Also decide to stop practicing one or more behaviors that put you or your family at risk.

Chapter 7 Review

Ideas to Remember

- Exploitation means using someone for selfish advantage without regard to the person's rights, needs, or feelings.

- Sexual activity can be positive or negative, depending on the intent of the people involved.

- Sexual harassment involves the use of power to force, or try to force, someone to engage in unwanted sexual behavior.

- Rape generally refers to forced sexual intercourse. Forcing anal intercourse on a man or woman also is commonly considered rape. Statutory rape is sexual intercourse with a minor.

- Rape can happen to anyone.

- Female adolescents and women in their early twenties are at the greatest risk for sexual assault.

- Most young women and adolescent girls who are raped know their attackers. In a dating situation, this is called date rape.

- The victim of sexual exploitation or rape is never at fault.

- The term incest usually refers to sexual activity between an adult and a child in a family.

- Sexual harassment, rape, and incest have more to do with abuse of power than with sexuality.

Questions to Ask Yourself

1) How would you define the following: sexual harassment, rape, incest?

2) Why are adolescent girls and women in their early twenties at high risk for being raped?

3) What can men do to help prevent themselves from raping women?

4) What are three things women can do to help prevent date rape?

5) What are three things men can do to prevent date rape?

Action for Health

EXAMINING YOUR GENDER ATTITUDES

The attitudes that boys and girls have toward one another quite often determine how they relate together. The attitudes about gender that a man and woman in a relationship have may affect the success of their relationship.

Work with your teacher on this activity to examine your own gender attitudes. The class should be divided in two groups: males and females. Both groups read the "Suggestions for Preventing Date Rape" on page 121. After reading, the girls comment about the first point listed under "Men." The boys then respond to the girls' comments. Similarly, the boys comment on the first point listed under "Women." Then the girls respond to the boys' comments. Examine your attitudes toward all the suggestions in this way. Have one person from each group summarize the discussion.

EXAMINE SOCIETY'S GENDER ATTITUDES

To explore society's attitudes toward women, look up information about:

- the types of jobs women have
- how much money women and men earn for the same jobs
- the role of women and men as homemakers
- how advertising and popular magazines portray the sexual role of men and women

Your local librarian can help with this activity. After collecting this information, write a summary about society's attitudes toward women. As a class, discuss how these attitudes might influence sexual exploitation.

Chapter 7 Review Sheet

Name _____ Date _____ Period _____

Part A Define each of the terms below in your own words.

1) Sexual harassment

_____
_____
_____
_____

2) Sexual exploitation

_____
_____
_____
_____

3) Rape

_____
_____
_____
_____

4) Incest

_____
_____
_____
_____

5) Date rape/acquaintance rape

_____
_____
_____
_____

Part B Use the definitions in Part A to help answer these questions.

6) Write an example of sexual harassment.

_____

_____

_____

_____

7) Explain the difference between sexual harassment and rape.

_____

_____

_____

_____

8) List three suggestions for preventing rape.

_____

_____

_____

_____

9) What do sexual harassment, rape, and incest have in common?

_____

_____

_____

_____

10) List three things a rape victim should do.

_____

_____

_____

_____

Chapter 7 Critical Thinking

Name _____ Date _____ Period _____

The excerpt below is from "Policy on Sexual Harassment of Students," a student handbook distributed in 1991 at a high school in Massachusetts. Read it carefully. Then answer the questions that follow.

Definition: Sexual harassment is unwanted sexual attention from anyone with whom the victim may interact in the course of receiving her/his education in school or at school-sponsored activities.

The following behaviors from a peer are considered sexual harassment:

- Staring or leering with sexual overtones
- Spreading sexual gossip
- Unwanted sexual comments
- Pressure for sexual activity
- Unwanted physical contact of a sexual nature

If you believe that you have been sexually harassed by another student or if you have questions about this issue, seek the help of an adult whom you trust, such as a teacher, guidance counselor, your parent/guardian, the dean of students, or an administrator. Any accusation of sexual harassment will be investigated by the dean of students, and a written record of the investigation will be filed with the principal.

If it is determined that a student has sexually harassed another student, the possible consequences for the offender are:

- Parent conference
- Apology to the victim
- Detention
- Suspension
- Recommendation for expulsion from school
- Referral to police

Reprisals, threats, or intimidation of the victim will be treated as serious offenses that could result in expulsion. In every case, a high degree of confidentiality will be maintained to protect both the victim and the offender. All efforts will be made to preserve the victim's sense of control in the situation.

Consider the excerpt you have just read and remember what you have learned from the chapter as you answer the questions below.

1) According to the "Policy on Sexual Harassment of Students," would a sexual joke that a student blurted out in class be sexual harassment? Use information from the excerpt to support your answer.

_____
_____

2) According to the policy, would a student who persists in asking for a date from an unresponsive party be guilty of sexual harassment? Support your answer with sentences from the excerpt.

_____
_____

3) What legal protections exist for a victim of sexual harassment? Do you think it is a good idea for schools to include their own sexual harassment guidelines? Explain your reasons.

_____
_____

4) Reread the consequences to the offender. Give an example of when you think an apology to the victim is appropriate. Give a second example of when you think a referral to the police is appropriate.

_____
_____

5) Why do you think the passage includes this sentence: "In every case, a high degree of confidentiality will be maintained to protect both the victim and the offender"?

_____
_____

Appendix A: Sexually Transmitted Diseases

Knowing the facts can prevent a person from getting an STD. For that reason, Appendix A provides information about some common STDs. For most STDs, signs or symptoms of infection may be absent or might disappear. Either way, those STDs remain in the body until a person receives treatment. A few STDs remain in the body forever and only the symptoms can be treated.

Here are some general signs of STDs:

Women:
- Any unusual discharge or smell from the vagina
- Pain in the pelvic area
- Burning or itching around the vagina
- Bleeding from the vagina that is not menstruation
- Pain deep inside the vagina during sexual intercourse

Men:
- Any drip or discharge from the penis

Either Women or Men:
- Sores, bumps, or blisters near the sex organs or mouth
- Burning and pain during urination or bowel movements
- Swelling or redness in the throat
- Flulike feelings (fever, chills, aches)
- Swelling in the groin

Because a person with an STD can give it to others, treatment is all the more important. To stop the spread of STDs, the wise decision is to be checked when a person suspects infection. Any of the specific signs of STDs listed in Appendix A can indicate the presence of an STD. In addition, sexually active people should get checked for STDs at their annual health exams.

Four of the top ten reported STDs in the United States are chlamydia, gonorrhea, syphilis, and AIDS. Table A.1 provides information on the first three. Information on AIDS is provided on page 135. Table A.2 lists other common STDs. The graphs show estimated new cases and costs of common sexually transmitted diseases in the United States each year.

Appendix A: Sexually Transmitted Diseases

Table A.1. Common Sexually Transmitted Diseases

	Chlamydia	Gonorrhea	Syphilis
Cause	Bacterium (*Chlamydia trachomatis*)	Bacterium (*Neisseria gonorrhoeae*)	Bacterium (*Treponema pallidum*)
Symptoms			
Males	Mild discharge from penis, painful urination, testicular pain; two out of four males have no symptoms	Discharge from penis, burning during urination; some men have no symptoms	First symptom is a chancre, or painless sore, on the body part exposed to the partner's sore; symptoms of second-stage disease are rash and flulike problems; patchy hair loss and swollen glands throughout body may also occur; imitates symptoms of other diseases
Females	Mild vaginal discharge, painful urination, painful intercourse, bleeding between periods, lower abdominal pain; two out of three females have no symptoms	Painful urination, yellow or bloody vaginal discharge, abdominal pain, bleeding between periods, vomiting, fever; many women have no symptoms	
Diagnosis	Test urine or cells from genital area	Usually one or more of three different tests of cells from genital area	Doctor's recognition of the symptoms, microscopic identification of bacteria, and blood tests
Treatment	Prescription antibiotics and other medicines	Prescription antibiotics alone or in combination	Shot of penicillin or other antibiotics
Complications if not treated			
Females	Pelvic inflammatory disease (PID), sterility; tubal pregnancies, which can be life threatening	PID; sterility; tubal pregnancies, which can be life threatening; increased risk of HIV infection	Increased risk of HIV infection; third stage can lead to damage to heart, eyes, brain, nervous system, bones, joints, or other body parts; mental illness; blindness; other neurologic problems; heart disease; death
Males	Rarely, pain or swelling in the scrotal area, a sign of epididymitis	Increased risk of HIV infection	
Newborns	Born with eye infection, pneumonia	Born with eye infection	Serious mental and physical problems; stillborn or die just after birth
Prevention	Regular tests of people with multiple sex partners, treatment for all who test positive, test of pregnant women, use of condoms or diaphragms, abstinence	Correct and consistent use of male condoms, test of pregnant women, treatment for all sex partners who test positive, abstinence	Avoiding contact with the open sores and other infected tissues, using condoms, screening and treatment of infected people, test of women early in pregnancy, abstinence

Table A.2. Other Sexually Transmitted Diseases

NAME	CAUSE	SYMPTOMS	TREATMENT	COMPLICATIONS	PREVENTION
Genital Herpes	Herpes simplex virus (HSV) type 2	Extremely painful clusters of small blisters in genital area, around anus, and on buttocks and thighs; painful urination; headache and flulike symptoms; blisters heal and recur at varying times	No known cure because the virus remains in certain nerve cells for life; drugs limit severity and length of episodes; careful hygiene and stress management help	Episodes can be long-lasting and severe; pregnant women can pass virus to fetus; babies are born with brain inflammation, severe rashes, eye problems	Avoid sexual contact until sores are healed; use condoms between outbreaks; limit sex partners; a new drug may possibly suppress viral activity and prevent recurrences
Genital Warts (condyloma acuminata or venereal warts)	A few types of human papillomavirus (HPV)	In women, warts occur, often in clusters, outside and inside vagina, on cervix or labia, around anus; some have no obvious symptoms. In men, warts are less common; if seen, warts occur on tip or shaft of penis, scrotum, around anus	Treatments eliminate warts but not the virus; removal with acid, a solution, prescribed creams, freezing, burning, laser or other surgery, or injection with a drug; warts can reappear after treatment	Virus remains in the body and can cause later problems such as precancerous cervical disease or cervical and other genital cancers; problems during pregnancy; mothers can transmit disease to fetus	Avoid contact with the virus; avoid sexual contact until warts are treated; use condoms, abstinence
Trichomoniasis	Parasite (*Trichomonas vaginalis*)	Often occurs without symptoms. In women, heavy yellow-green or gray vaginal discharge, vaginal odor, painful urination, irritation and itching of genital area. In men, thin, whitish discharge from penis and painful or difficult urination	Single dose of a drug given to both partners	Increased risk of transmitting HIV; women can deliver low birth-weight or premature babies	Use of male condoms, abstinence
Pubic Lice	Tiny parasitic insects (pediculosis pubis or crab lice) that infest pubic hair and feed on human blood	Itching in the pubic area	Prescription and over-the-counter lotions, creams, and shampoos	Scratching may spread lice to other parts of body	Treating all people with whom person has come in contact, cleaning or washing all clothing and bedding at high temperatures

Cases of Common Sexually Transmitted Diseases

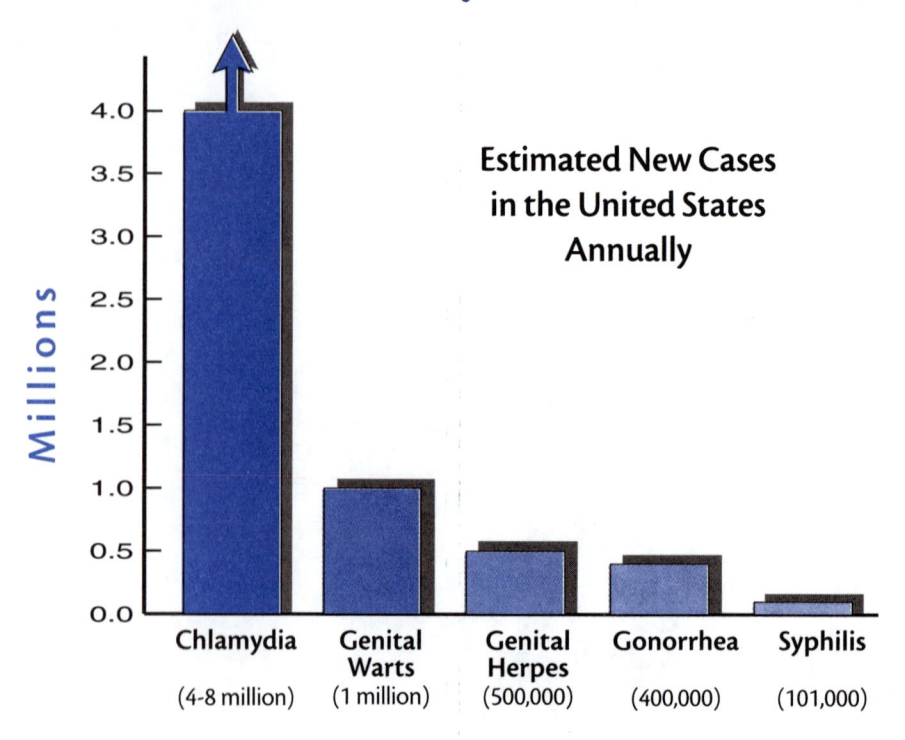

Estimated New Cases in the United States Annually

Chlamydia	4-8 million
Genital Warts	1 million
Genital Herpes	500,000
Gonorrhea	400,000
Syphilis	101,000

Source: National Institute of Allergy and Infectious Diseases, 1998

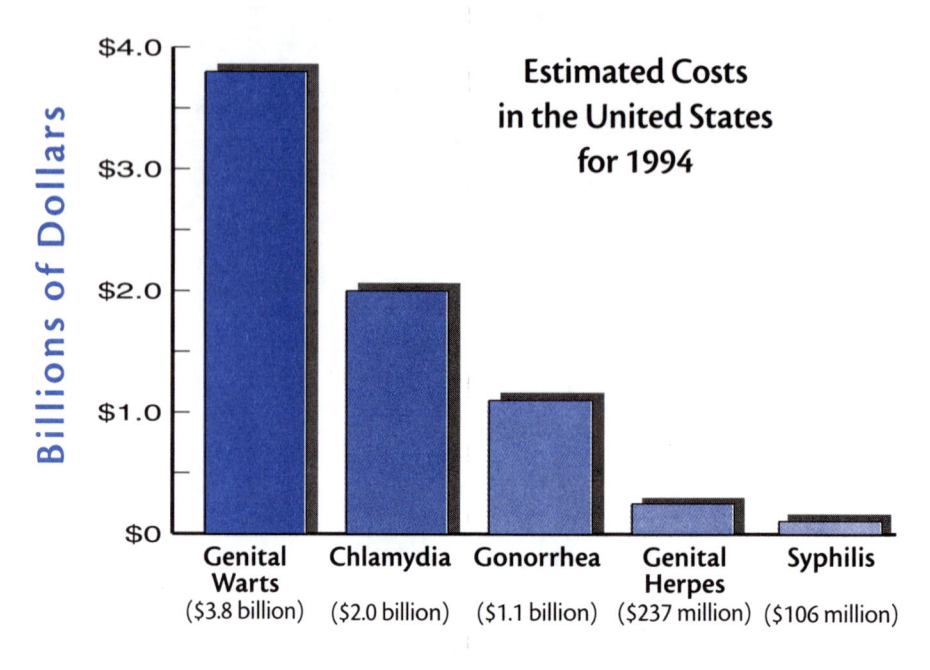

Estimated Costs in the United States for 1994

Genital Warts	$3.8 billion
Chlamydia	$2.0 billion
Gonorrhea	$1.1 billion
Genital Herpes	$237 million
Syphilis	$106 million

Source: T.R. Eng & W. T. Butler, eds. *The Hidden Epidemic: Confronting Sexually Transmitted Diseases.* Washington, DC: National Academy Press, 1997

HIV/AIDS Facts

Figures
About 30.6 million people living with HIV/AIDS and about 11.7 million cumulative AIDS-related deaths worldwide through December 1997
As many as 900,000 Americans infected in 1997
AIDS is six times higher among African Americans and three times higher among Hispanics than among whites
One-quarter to one-third of untreated infected pregnant women pass the infection to their babies

Cost
Cost of sexually transmitted HIV infection in the United States totaled $6.7 billion in 1994

How HIV/AIDS Is Acquired
Through exchange of body fluids—blood, semen, vaginal secretions, breast milk—with an infected person
Unprotected sexual contact with an infected partner
Sharing needles contaminated with HIV-infected blood among injection drug users
Mother to child before, during, or after birth or when baby drinks mother's breast milk

How HIV/AIDS Is NOT Acquired
Contact with saliva, sweat, tears, urine, or feces
Casual contact such as kissing, hugging, shaking hands, crying, coughing, or sneezing
Sharing food utensils, towels and bedding, swimming pools, telephones, or toilet seats
Biting insects such as mosquitoes or bedbugs

Early Symptoms of HIV Infection
Some people have no symptoms
Others have a flulike illness (fever, headache, sickliness, swollen lymph glands) within a month or two that disappears in a week to a month
More severe symptoms may not appear for up to ten or more years in adults and within two years in children born with HIV

AIDS Symptoms
After immune system is weakened, constantly enlarged lymph nodes, lack of energy, frequent fevers and sweats, persistent skin rashes
Opportunistic infections that cause such symptoms as coughing, shortness of breath, seizures, persistent diarrhea, vision loss, extreme fatigue
Difficult-to-treat cancers such as Kaposi's sarcoma or lymphomas

Diagnosis
Blood test for antibodies to HIV

Treatment
No known cure, but drugs with serious side effects fight HIV infection and its associated infections and cancers
Radiation, chemotherapy, or injections of alpha interferon for Kaposi's sarcoma and other cancers

Prevention
Avoid unprotected sex and sharing needles among injection drug users
Use latex condoms during vaginal, oral, or anal sex

Appendix B: Events in Pregnancy

CHANGES DURING PREGNANCY

Trimester/Month	Embryo/Fetus	Mother
First Trimester		
First Month	Embryo 1/4 inch long; heart is obvious; eyes, nose, and brain appear; arms and legs are small bumps	Menstruation stops; breasts are tender; fatigue; frequent urination; positive pregnancy test
Second Month	Embryo 1 inch long; head quite large; fingers, toes appear; nervous system and brain coordinate body functions; mouth opens and closes	Possible nausea and vomiting lasting into third or fourth month
Third Month	Embryo becomes a fetus; 3 inches long; weighs about 1/2 ounce	Breasts possibly swollen; thickened waist; uterus well rounded
Second Trimester		
Fourth Month	Fetus 6 inches long; weighs 3 1/2 ounces; sex may be determined; body covered with soft hair	Breasts may discharge liquid; uterus wall more stretched; larger midsection
Fifth Month	Fetus 9 1/2 inches long; weighs 10 ounces; skin less transparent	Feels stronger fetal movements; uterus higher; breasts not much larger; midpoint of pregnancy
Sixth Month	Fetus 12 inches long; weighs 1 1/2 pounds; eyebrows, eyelashes showing; sucks thumb; hiccups; if born at this time, usually dies	Uterus increases in size; fetal movements may feel sharp
Third Trimester		
Seventh Month	Fetus 14 inches long; weighs 2 1/2 pounds; covered with a greasy substance to protect it; responsive to sound and taste; if born at this time, moves actively, cries weakly; can live with expert care	Still feels fetal movements; size of uterus increases; weight gain continues
Eighth Month	Fetus 15 1/2 inches long; weighs 3 1/2 pounds; hair on head; skin red and wrinkled; if born, can survive with proper care	Gains some additional weight; uterus extends; fetal movements continue
Ninth Month	Fetus 18 inches long; weighs 5 1/2 pounds; gestation period ends; body fat makes figure more round, less wrinkled in face	Top of uterus nears breastbone; possible frequent urination because of pressure on bladder; possible difficulty in walking

9 weeks

11 weeks

15 weeks

36 weeks

FETAL DEVELOPMENT

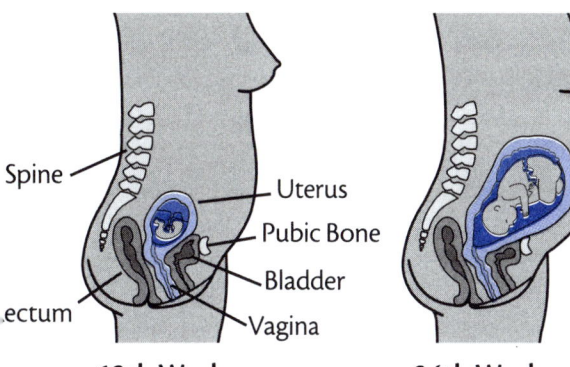

12th Week
Fetus weighs 0.4 oz.

24th Week
Fetus weighs 1.5 lb.

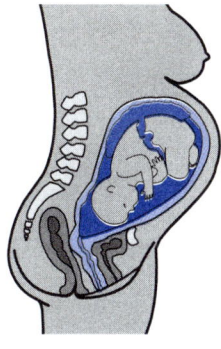

36th Week
Fetus weighs 7 lb.

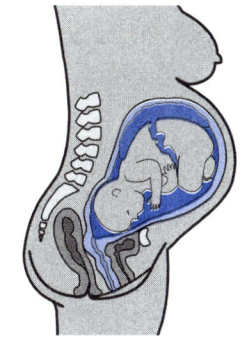

40th Week
(after lightening)
Fetus weighs 7.5 lb.

Labels: Spine, Uterus, Pubic Bone, Bladder, Vagina, Rectum

STAGES OF LABOR

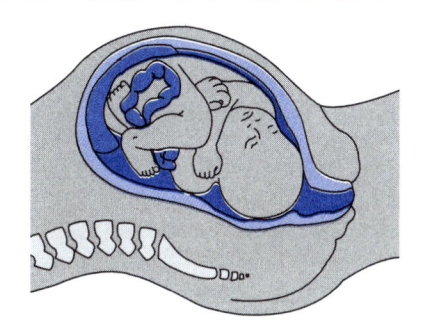

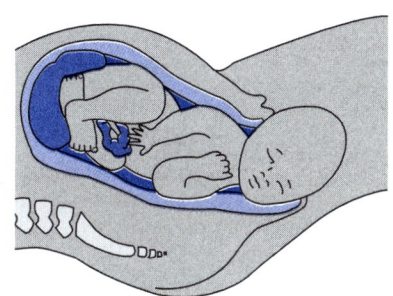

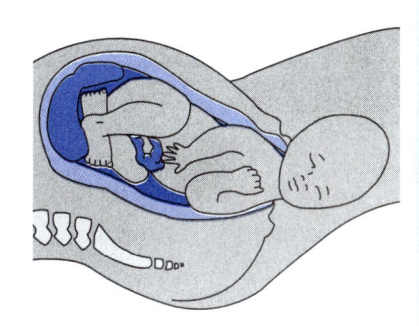

The process of childbirth has three stages. The **first stage of labor** begins when the uterus contracts to dilate, or open, the cervix to push the baby down and out of the vagina. The woman feels pains with each contraction, which become more and more frequent. At some time during this stage, the protective fluid surrounding the fetus breaks and flows out of the woman's body. The first stage of labor is much longer for a woman having her first baby (6 to 18 hours) than for one who has already had a baby (2 to 10 hours).

The **second stage of labor** starts when the cervix is fully dilated. The mother feels the desire to bear down with each contraction. She actually pushes the baby out. This stage ends with the birth of the baby. On average, this stage takes 50 minutes for first-time mothers and 20 minutes for women who have had a baby before.

Generally the baby's head appears first. The appearance of the head is called **crowning**. To prevent the mother's skin between the vagina and the anus from tearing during crowning, the doctor may make an incision called an **episiotomy**. A woman should talk with her physician before delivery about whether an episiotomy is necessary. If an episiotomy is done, the doctor sews the skin back together after delivery.

The **third stage of labor**, or afterbirth, occurs when the placenta is expelled from the uterus. This stage may last only a few minutes to as long as thirty minutes.

Some doctors consider the hour immediately after delivery to be a fourth stage of labor. That is a time when they watch the mother closely to make certain she is all right.

Appendix B: Events in Pregnancy

Glossary

A

Abortion—The ending of a pregnancy (p. 97)

Aggressiveness—Standing up for your own rights without regard for others' rights (p. 81)

Acquired immunodeficiency syndrome (AIDS)—A disorder of the immune system (p. 11)

Androgyny—The state of having what society defines as both masculine and feminine traits (p. 49)

Assertiveness—The ability to stand up for your own rights without violating another person's rights (p. 81)

C

Cervix—The narrowed part, or neck, of the uterus. (p. 27)

Chancre—A painless sore that is the first symptom of syphilis (p. 132)

Chlamydia—A sexually transmitted disease with few symptoms (p. 132)

Circumcision—A surgical procedure involving removal of the foreskin from the head of the penis (p. 30)

Clitoris—A tiny organ at the top of the innermost labia containing many nerve endings (p. 27)

Conception—The union of an egg and a sperm, causing pregnancy (p. 8)

Contraceptive method—The technique used to prevent the joining of an egg and a sperm (p. 8)

Crowning—The appearance of the baby's head during the second stage of labor in childbirth (p. 137)

D

Date rape—A form of rape in which the victim has been dating the attacker (p. 119)

Diaphragm—A contraceptive device that fits over the cervix, working as a barrier to sperm (p. 103)

Douching—Flushing the vagina with liquid; sometimes used as a contraceptive method (p. 99)

E

Ejaculatory duct—A duct extending through the center of the prostate gland to the urethra (p. 30)

Endocrine system—The body system consisting of ductless glands that release hormones (p. 22)

Episiotomy—An incision between the vagina and the anus that prevents the mother's skin from tearing during childbirth (p. 137)

Estrogen—A female sex hormone (p. 23)

F

Fallopian tubes—Two tubes, one extending from each ovary to the uterus, in which released eggs travel (p. 25)

Female condom—A prelubricated latex contraceptive device that completely lines the vagina to prevent sperm and sexually transmitted diseases from entering (p. 100)

First stage of labor—The stage of childbirth in which the uterus contracts to push the baby down and out of the vagina (p. 137)

Foreskin—A fold of skin covering the head of the penis (p. 30)

G

Gender identity—The self-image of being either male or female (p. 40)

Gender role stereotyping—Confining expectations of how males or females should act (p. 40)

Genital herpes—A sexually transmitted disease caused by the herpes simplex virus type 2 (p. 133)

Genital warts—A sexually transmitted disease caused by a few types of human papillomavius (p. 133)

Gonorrhea—A communicable disease caused by a bacterium that usually is transmitted sexually (p. 132)

H

High-risk behavior—Actions or activities that place a person in greater than average danger (p. 58)

Human immunodeficiency virus (HIV)—The pathogen that causes AIDS (p. 11)

Hymen—A thin membrane inside the vaginal opening (p. 27)

Hypothalamus—A small area in the brain that sends a message to the pituitary gland to release certain hormones during puberty (p. 22)

I

I statement—A way to share feelings about something that you want to resolve without putting the other person down (p. 82)

Incest—Sexual activity between family members, usually involving a child and an adult (p. 123)

Intrauterine device (IUD)—A small, T-shaped object inserted into the uterus to disrupt the normal chemistry of the uterine lining (p. 103)

L

Labia—Two folds of skin surrounding the vaginal opening (p. 27)

M

Male condom—A thin sheath, usually made of latex rubber, that is worn over the penis to stop sperm from entering the vagina (p. 100)

Masturbation—Self-touching or self-stimulation of the genitals for sexual pleasure (p. 9)

Mutuality—Having the same feelings for each other (p. 78)

N

Natural family planning—A series of conception control methods based on determining when a woman ovulates (p. 102)

Nocturnal orgasm—A person's sexual arousal and response that occurs during sleep (p. 30)

Nonassertive—Giving up your rights so that other people can do whatever they want (p. 81)

Nonverbal communication—Body language (p. 87)

O

Oral contraceptives—Pills taken daily by mouth to prevent conception (p. 103)

P

Pituitary gland—A small gland in the brain that releases hormones that influence many other glands; the master gland (p. 22)

Prostate gland—A small muscular gland that produces a neutralizing fluid that protects sperm and aids in sperm movement (p. 30)

Pubic lice—A sexually transmitted problem caused by direct contact with a parasitic insect that infests pubic hair (p. 133)

R

Rape—Sexual intercourse without consent and with actual or threatened force (p. 118)

Reciprocity—Mutual sharing or exchange of information between two people (p. 78)

S

Scrotum—The sac-like structure holding the testes (p. 29)

Second stage of labor—The stage of childbirth in which cervix is fully dilated and the baby is pushed out (p. 137)

Self-disclosure—Revealing your real self and honest feelings to other people (p. 74)

Seminal fluid—A fluid secreted by the seminal vesicles to mix with sperm (p. 29)

Seminal vesicle—One of two glandular structures that secretes seminal fluid during ejaculation (p. 29)

Sexual exploitation—Taking advantage of another person using sexual activity as a weapon (p. 116)

Sexual harassment—Exploitation using power or status to force someone into sexual behavior (p. 116)

Sexuality—A personal and natural expression of a person's identity beginning at birth and continuing until death (p. 6)

Sexual responsibility—The ability to speak and act honestly and directly, and to show respect and care in a relationship (p. 58)

Sexually transmitted disease (STD)—Any disease that is spread through sexual activity (p. 8)

Sexual orientation—A person's gender preference regarding sexual relations (p. 42)

Sterilization—A surgical procedure that makes a person incapable of conception (p. 105)

Syphilis—An infectious disease caused by a bacterium that is usually transmitted sexually (p. 132)

T

Testosterone—The male sex hormone (p. 22)

Third stage of labor—The stage of childbirth in which the placenta is expelled from the uterus; afterbirth (p. 137)

Toxic shock syndrome (TSS)—An infection caused by bacteria present in the vagina during menstruation (p. 102)

Tubal sterilization—A sterilization procedure for women that blocks the fallopian tubes (p. 105)

U

Urethra—The passageway from the bladder to the opening of the penis through which urine and semen pass, each at different times (p. 30)

V

Vaginal spermicide—A contraceptive such as a foam, cream, jelly, suppository, or sponge that is inserted into the vagina to destroy sperm (p. 101)

Vasectomy—A sterilization procedure for men that blocks the passage of sperm (p. 105)

W

Withdrawal—An unreliable contraceptive method involving the removal of the penis from the vagina before ejaculation (p. 99)

Index

A

"Abby, Dear," 113-14
Abortion, 107
    adolescent pregnancy and, 97-98
    defined, 97
Abstinence, 11, 59, 68, 97-98
Acne, 6, 23
Action for Health, 14, 34, 52, 68, 90, 110, 126
Active listening, 81-82
Adolescence, 19-31
    changes during, 20-21
    communicating with parents, 75-78
    defined, 20
    marriage during, 63
    physical changes during, 22-30
    pregnancy/parenthood during, 10, 12, 57, 63, 96-97
    responsibility and, 21-22, 31
    sexually transmitted diseases and, 64, 97
Adoption, 98
Adultery, portrayed on television, 17
Aggression
    gender and, 41, 48
    in a relationship, 81
AIDS, 11, 20, 59, 98, 131-35. *See also* Human immunodeficiency virus
Aid to Families With Dependent Children, 96
Alcohol, 8-9, 119-21
Alfano, Kathleen, 55
Alternatives, when saying "no," 84
American Medical Association (AMA), 10
Anal intercourse, 118
Androgyny, 49
Anger, 86
Animals and human behavior, 71-72
Arguments. *See* Conflict, resolving
Assertiveness
    date rape and, 120-21
    gender and, 41
    keeping limits and, 65, 84
    in a relationship, 81-84, 86
Assumptions, 83

B

Behaviors
    gender-related, 40-42
    high-risk, 58-59
    requesting change in, 86
    and sexual responsibility, 58-60
Bernhard, Yetta, 21
"Bernice Bobs Her Hair," 93
Birth process, 25-26
Birth canal. *See* Vagina
Birth control pills, 8, 102-03
Bisexuality, 42
Body hair, 23-24
Body image, 60-62
Body language, 81, 87
Boisterousness, gender and, 41
Boundaries. *See* Setting limits
Breast
    cancer, 8, 31
    development, 23-24
    self-examination, 8, 31, 34

C

Calendar method, 102
Cancer
    breast, 8, 31
    cervical, 31
    testicular, 31
Cervix, 27, 103, 137
    cancer of, 31
Child abuse. *See* Incest; Sexual abuse/assault
Chlamydia, 132
Circumcision, 30
Clear messages, 84
Clitoris, 27
Communication, 73-87
    and assertiveness, 81-83
    and criticism, 85
    date rape and, 120-21
    "do's" and "don'ts," 82-83
    with friends, 78-80
    good, 81
    limits and, 76-77
    nonverbal, 87
    with parents, 75-78
    resolving conflict, 85-86, 90
    saying "no," 84
    and sexuality, 75
    touch, 47-49, 79-80
    verbal, 76, 121
Complaints, in an argument, 83, 85-86
Compromise, 62, 82
Conception, 8
    issues regarding, 62-63
    process of, 25. *See also* Contraception; Pregnancy
Condoms, 99, 100
Conflict, resolving, 85-86, 90
Consequences of actions, considering, 62-63
Contraception, 11
    and diabetes, 103
    issues regarding, 62-63, 95
    talking to parents about, 77
    types of, 97-106
Contraceptive methods, 8, 97
    abstinence, 97-98
    considerations when choosing, 106-07
    nonprescription, 99-102
    prescription, 102-04
    sterilization, 104-05. *See also* Names of specific methods
Contraceptive sponge, 101
Counselors, seeking help from, 65, 110, 117, 123
Courtesies, with parents, 77-78
Creativity and gender, 41
Crowning, 137
Criticism, handling, 85
Cultural values, 63

D

Dan, Alice, 37
Date rape, 57, 119-21
Dating, 61
    double, 64
    gender roles in, 44-45
    setting limits in, 61, 84
"Dear Abby," 113-14

Decision making
  as a couple, 57
  developing, 62-63
  responsible, 97
  steps of, 57, 62-63
Defensiveness, 82, 85
Diabetes and contraceptives, 103
Diaphragm, 103
Dilation of cervix, 27
Dominance and gender, 41
Double dating, 64
Douching, 99
Drugs, 120-21
Ductless glands, 22

E

Education, 12, 65
Eggs, 22, 24-26, 28, 105
Ejaculation, 8, 23, 29, 99
Ejaculatory duct, 30
Emotions
  development of, during adolescence, 11
  gender and, 49
  expressing, 74, 117
Endocrine system, 22
Entertainment Machine, The, 17
Episiotomy, 137
Erection, 9, 24, 30
Estrogen, 23, 28, 102
Exploitation, 116. *See also* Sexual abuse/assault
Eye contact, 82

F

Facial hair, 24
Fallopian tubes, 25, 99, 105
Family
  communication in, 74
  incest in, 123
  learning about love from, 62
Family planning
  contraception and, 97-107
  gender roles and, 46
  importance of, 96-97
  sexual responsibility and, 58-62. *See also* Contraception; Natural family planning
Feedback
  asking for, 86
  giving, 81-82, 85

Feelings
  expressing, 86
  sexual responsibility and, 58
  sharing, 82
Female condom, 100
Feminine traits/characteristics, 39, 49
  parental influence and, 43
  stereotypes of, 43-44
Fetal development, 137
Fertilization, 26. *See also* Contraception, types of
Fitzgerald, F. Scott, 93
Foam, as vaginal spermicide, 100
Fondling, 47-49, 61
Food, 8
Forced intercourse. *See* Rape
Foreskin, 30
Friendship
  communication and, 78-79
  developing, steps to, 78-79
  gender roles in, 49
  versus intimate relationship, 79-80

G

Games, playing to manipulate, 61, 83
Gay. *See* Homosexuality
Gender
  attitudes, 126
  characteristics, 39
  identity, 40, 60
Gender roles, 40
  changing, 44-45
  defined, 40
  effect on relationships, 44-45
  effect on touch and communication, 47-49
  effect on you, 42
  expectations, 52
  future of, 49
  healthy, 42, 49
  learning about, 43-44
  and sexual orientation, 42
  stereotypes of, 40-42
Genitals, self-touch of, 47. *See also* Masturbation; Reproductive system; Names of specific organs
Genital herpes, 133
Genital warts (HPV), 64, 133
Gonorrhea, 132
Group dating, 64

Growth spurt, 22
Guilt and sexual behavior, 58, 60

H

Health issues and choosing a contraceptive, 107
Herpes, 133
Heterosexuality, 42
High-risk behaviors, 58-59
HIV (human immunodeficiency virus), 11, 100, 135. *See also* AIDS
Holding hands, 61
Homosexuality, 42
Honesty, 46, 48, 58, 65, 82. *See also* Self-disclosure
Hormones, 22-23, 102-03
  sex, 23, 28
Hugs, 49
Hymen, 27
Hypothalamus, 22

I

"I count," and sexual responsibility, 65
Identity
  discovering, 20-22, 40, 45
  sexual, 6
Impatience, 83
Incest, 123
  reporting, 65
Intellectual development, 20
Intelligence and gender, 41
Intercourse. *See* Sexual intercourse
Intimacy, differences from friendship or sexual intercourse, 79-80
Intrauterine device (IUD), 102-03
"I statements," 82, 86
IUD (intrauterine device), 102-03

K

Kissing, 49, 61, 116

L

Labeling, 40
  poor communication and, 83
  sexual preference and, 42
Labia, 27
Labor, stages of, 137
Latex condoms, 99-100
Lifestyle, contraception choices and, 106

Limits, setting sexual, 60-65, 68
"Lines," 48, 63-64
Listening, 81-82
Love
    giving and receiving in a relationship, 62
    sex and, 60
Lubricant, condoms and, 99-100

M

Magazines, 7, 48-49
Manipulating, 61, 83
Marriage
    gender roles in, 48-49
    manuals, 48
    teenage, failure rates of, 63
Masculine traits/characteristics, 39, 49
    and parental influence, 43
Master gland. See Pituitary gland
Masturbation, 9, 47
    myths regarding, 9, 47, 60
Maturity and sexual behavior, 60-62
Menses. See Menstruation
Menstruation, 23, 26
    contraception and, 101
    cycle, 27-28
Miscarriage, 107
Misunderstandings, preventing, 85
Mixed messages, 84, 87
Moral development, 21
Mucous, vaginal, 27
Mutuality, 28
Myths and misconceptions, 7-9
    about masturbation, 47, 60
    about rape, 118

N

Name-calling, 83
National health goals, 10-12
Natural family planning, 102
Nocturnal orgasm, 30
Nonassertive behavior, 81
Nonprescription contraceptive methods, 99-102
    abstinence, 11, 59, 68, 97-98
    condom, 99-100
    contraceptive sponge, 101
    douching, 99
    natural family planning, 102
    spermicidal creams and jellies, 101, 104
    spermicidal suppositories, 101
    vaginal spermicides, 100-01
    withdrawal, 99
Nonverbal communication, 87, 90, 121. See also Body language

O

Old issues, 83
Open-ended questions, 83, 86
Oral contraceptives, 102-03
Orgasm. See Nocturnal orgasm
Ovaries, 23-25, 28
Ovulation, 23, 25, 28, 102

P

Pap smear, 31
Parents
    communicating with, 65, 75-78
    gender roles and, 43
    influence on sexual attitudes, 9, 19, 47, 62-64
    seeking advice from, 5, 7, 12
Passive behavior. See Nonassertive behavior
Peers
    gender roles and, 43
    influences of, 20, 43
    pressure from, 57, 63-65
Pelvic examination, 31
Penis, 30, 99-100
    discharge from, 8
    size of, 9, 30
Period. See Menstruation
Perspiration, 23
Physical activity, 9
Physical development during adolescence, 22, 41
Physical disclosure and false intimacy, 98
Pill, the. See Oral contraceptives
Pituitary gland, 22
Post-ovulatory phase, 28
Pregnancy, 25-27, 136-37, 116-17
    adolescence and, 10, 96-97
    myths regarding, 8-9
    preventing, 11. See also Contraception
Premenstrual phase, 28
Pre-ovulatory phase, 28
Prescription contraceptive methods, 102-04
    diaphragm, 103
    IUD, 103-04
    oral contraceptives, 102-03
Primary sex characteristics, 23
Printed materials, and gender roles, 44
Private space, 61
Pro-choice advocates, 107
Progesterone, 102-04
Pro-life supporters, 107
Prostate gland, 30
    examination, 31
Puberty, 22-23. See also Physical development
Pubic hair, 24
    lice, 133

R

Rape, 11, 118-19
    alcohol and, 9
    date-, 57, 119-21
    defined, 118
    male victims of, 9
    myths about, 118
    preventing, 120-21
    reporting, 65, 122-23
Reciprocity, 78
Rejection, 6, 845
Religious values, 63-64, 110
Reproduction, becoming capable of, 22-30
Reproductive system
    female, 22-28
    male, 22, 28-30
Respect, in a relationship, 46, 58, 82
Responsibilities, 21, 31, 65
Rhythm method, 102
Roles, 39-49
    changing, 44-45
    gender, 41-49
    traditional, 43-48

S

Safe days, 102
Sarcasm and communication, 83
Saying "no," four-step plan, 84
Scrotum, 29, 104
Secondary sex characteristics, 23
Self-concept and sexual choices, 6, 64
Self-discipline, enforcing limits and, 65

Self-disclosure
   communication and, 74
   criticism and, 85
   friendship and, 78-79
   intimacy and, 79, 81
   marriage and, 79
Self-esteem
   gender and, 41
   sexuality and, 6
Self-stimulation. See Masturbation
Semen, 30, 100
Seminal fluid, 29
Seminal vesicles, 29
Setting limits, 60-65
   reasons for, 68
Sex, defined, 6
   resistance to, 48
Sex roles. See Gender roles
Sexual abuse/assault, 9
   harassment, 116-18
   incest, 123
   rape, 118, 120-21
   reporting, 65
Sexual desire
   changes in during adolescence, 23-24
   myths regarding, 8
Sexual exploitation, 116. See also Sexual abuse/assault
Sexual harassment, 116-18
   reporting, 117
Sexual identity. See Gender identity
Sexual intercourse. See also Sexual abuse/assault
   abstinence from, 11, 59, 68, 97-98
   anal, 118
   consequences of, 58
   contraception and, 97-106
   myths about, 7-9
   pressure to participate in, 57, 63-65
Sexuality, 5-12
   adolescents and, 10-12
   communication and, 75
   defining, 6
Sexually transmitted diseases (STDs), 8, 11, 57, 97, 131-35
   controlling, 100, 107
   reporting, 65
Sexual organs. See also Names of specific organs

female, 24-28
male, 22, 28-30
Sexual orientation, 42, 60
Sexual preference, 42
Sexual relations. See Sexual intercourse
Sexual responsibility, 57-65
   decision making and, 62-63
   defined, 58
   limits and, 63-65
   pathway to, 59-62
Skin, changes in during adolescence, 23
Skin condoms, 100
Social development during adolescence, 20
Social skills and gender, 41
Sperm, 9, 28-30, 99, 104
   production of, 23
Spermicides, 100-01, 103
   creams and jellies, 101, 104
   suppositories, 101
Sponge. See Contraceptive sponge
Stages of labor, first, second, and third, 137
Statutory rape, 118
Sterilization, 47, 104-05
Support for victims of sexual assault, 117
Sweat glands, changes in during puberty, 23
Syphilis, 132

T

Tampons and the hymen, 27
Teachers, seeking help from, 65
Teenagers. See Adolescence
Television, 7
Testosterone, 22, 29
Testes, 23, 29
Testicles
   cancer of, 31
   self-examination of, 31
Time limits, 64
Touch
   communication, 47-49, 79-80
   gender roles and, 41
   setting limits on, 61
   sexual. See Fondling
   sexual assault and. See Sexual abuse/assault
Toxic shock syndrome (TSS), 101

Toys, and gender stereotyping, 55
Traits, gender-related, 31, 40-41
Trichomoniasis, 133
Trust
   in friendship, 79
   and incest, 123
Tubal sterilization, 105

U

United States Public Health Service (USPHS), 10
Urethra, 9, 29-30
Urine, 9, 30
Uterus, 24-28, 104

V

Vagina, 9, 24, 27, 99-100, 103
   irritation in, 31
   lips of, 27
   opening of, 27, 99
Vaginal spermicides, 100-01
Values
   decision making and, 63-64
   sexuality and, 5-6, 9
Vas deferens, 104
Vasectomy, 104
Verbal communication. See Communication
Verbal skills and gender, 41
Violent crimes. See Rape
Voice, change in during adolescence, 23-24

W

"Wet dreams," 30. See also Nocturnal orgasm
Withdrawal, 99
Woman's right to choose, 107
Womb. See Uterus

X

X-rated movies, 60

Y

Yes-no questions, 83
"You statements," 82